Science Educators Under the Stars

Amateur Astronomers Engaged in Education and Public Outreach

Edited by:

Michael G. Gibbs
Marni Berendsen
Martin Storksdieck

© Copyright 2007 Astronomical Society of the Pacific
390 Ashton Avenue, San Francisco, CA 94112-1722

Printed by Odyssey Press, Inc.

First published 2007

Cover design and book layout: Leslie Proudfit
Cover image: Supernova Remnant E0102 in the Small Magellanic Cloud
Cover image credit: NASA, ESA, and the Hubble Heritage Team (STScI/AURA)

Library of Congress Control Number: 2007931719
ISBN 978-1-58381-315-7

Please contact address below for information on ordering books:
Astronomical Society of the Pacific
390 Ashton Avenue
San Francisco, CA 94112-1722, USA

Phone: 415-337-1100
Fax: 415-337-5205

book@astrosociety.org

Contents

Foreword

Terry Mann, President of the Astronomical League

For most of us it started on one of those long clear evenings. You look up at the sky and wonder what that bright star is over there or think, "WOW! Look at the Milky Way." It is just that easy to become hooked on amateur astronomy.

One great thing about being an amateur astronomer is you can make the hobby as easy or as challenging as you want. The more you study the sky, the more you will see. Seeing the Milky Way arching over your head makes the sky look endless. That is when the sense of awe will hit you. Suddenly, you feel a little small in the scheme of things.

You want to jump in a little deeper. You might join an astronomy club and learn more about the night sky. Before you know it, you will want to share your hobby with others. Nothing compares to watching someone look through the telescope for the first time. This new observer is likely to start asking questions. They want to know what that bright star was in the west last night. They want to understand the phases of the moon or why the seasons happen. That is when you know you are hooked. Others are interested in your hobby too and you are having as much fun explaining the sky as they are listening. It is the first time you have taken astronomy to the people. You cannot wait until the next event.

Right after I was asked to write this forward, I went to my astronomy club and asked the audience, "What is an amateur astronomer?" Here is some of what I heard. Amateurs enjoy the passion of sharing the sky. Astronomy is like my second childhood, something exciting is always happening. The amateurs are like a family, they are old friends who are always there. Amateurs contribute by doing real science. Some of the words I heard were: "enthusiastic," "beginning," "wonder," "excel," "educators," "photographers," "curious," "hobby." We all agreed: we enjoy astronomy and we love to share our knowledge.

In the past, the efforts of amateur astronomers to be informal science educators have been largely at the individual or astronomy club level. Little support for their outreach came from outside a person's astronomy club. Little was known about the experiences and practices of these dedicated people or what they needed to support their outreach. How extensive is public outreach among amateur astronomers and what kinds of outreach do they do? How knowledgeable are they about astronomy and how effective as educators? How valuable are their contributions to increasing the public's appreciation and interest in science? What would provide the most assistance to them?

In recent years we have seen increasing support and recognition of amateur astronomers for their involvement with the public through such organizations and programs as:

- Astronomical Society of the Pacific (www.astrosociety.org)
- Astronomical League (www.astroleague.org)
- Night Sky Network (nightsky.jpl.nasa.gov)
- Project ASTRO (www.astrosociety.org/education/astro/project_astro.html)
- Solar System Ambassadors (www2.jpl.nasa.gov/ambassador/)
- 4M Community (www.meade4m.com)

Even magazine publishers, such as *Astronomy* and *Sky & Telescope*, issue annual awards for outreach.

For the first time, this book provides a compilation of the recent research on amateur astronomy outreach and discussions on the personal experience of sharing the universe with others.

The first chapter is about the "spark." Many of us know that feeling. That is why sharing our knowledge is so important. We want others to experience that spark and watch our hobby grow.

Chapter 2 discusses how amateur astronomers' interest in sharing with others fits into the larger context of amateur astronomy pursuits.

Chapter 3 introduces the arena of "free-choice" learning and how amateur astronomers can best contribute to the education of students in and out of the classroom. Many schools do not have the funds needed for astronomy equipment and many teachers appreciate help with the required astronomy curriculum.

Chapters 4 though 7 summarize the current body of research about amateur astronomers and the public outreach they do. Chapters 4, 5, and 6 cover such topics as the practices of amateur astronomers regarding outreach, what they feel they need to support their efforts, their level of knowledge about astronomy, and their impact as educators. Chapter 7 reports on research into the role of women in amateur astronomy. Based on the research, each of these chapters offers specific recommendations for supporting and expanding the cadre of amateur astronomers involved in outreach.

Chapter 8 gives the research a voice as we hear from four amateur astronomers who share their personal experiences in their often touching and inspiring encounters with the public.

The last chapter opens the door to the future of amateur astronomy outreach: what are the next steps, how can communities and organizations utilize and improve support for these dedicated, knowledgeable, and enthusiastic individuals.

Amateur astronomers are an important part of science education and public outreach. New discoveries continually make our hobby exciting. Sharing the excitement passes on the spark of curiosity which inspires the desire to explore further.

1

The Amateur Spark

David Levy and Jim Kaler

Why and how does someone become an amateur astronomer? In almost every case, the answer involves a sudden spark. First, though, what on Earth are professional astronomers doing writing such a chapter? Doesn't everyone know that professionals can't tell the difference between Orion and the Dippers, let alone that between Jupiter and Saturn? And forget those between Hydra and Hydrus, M 31 and M 57, or a catadioptric reflector and an apochromatic refractor. Shouldn't a dedicated amateur—one who "loves the subject"—be writing this? The answer comes from one of the 20th century's greatest astronomers, an expert on the Milky Way, Bart Bok. "I began as an amateur" he once said, "then I became a professional astronomer, and in retirement I am an amateur again."

THE SPARK

The simplest things can change a life, can provide a life. For David, it was a two-second burst of a faint meteor, a celestial firework seen on July 4, 1956; for Jim, it was the sight of a pretty star a decade earlier. Stars have points, right? Just look at the gold ones on your last homework. Walking home from the store, his Grandma, who was unschooled except in life, said no they don't. And then in a perfect scientific moment, said "look up and see for yourself." The boy looked, and no they didn't have points. The star gliding above his head was colored, was orange. How remarkable! At that moment, Jim became an astronomer, and never looked back. For both of us, the fascination with the sky came from a spark generated by simply

4

looking up, from which came a deep sense of seeing a combination of mystery and beauty.

David didn't do much with his meteor at first, except to store the memory away for a few years. Then, during the summer of 1960, a broken arm from a bicycle accident led to the gift of a book called *Our Sun and the Worlds around It*. By the end of that summer, David was completely hooked. Becoming an astronomer was all he ever wanted to do. At first, his parents fretted that his all-consuming passion would affect his abilities at socialization and for making new friends. But as the years passed, his energy unabated, David's work resulted in plenty of them.

Jim's road was oddly similar. A trip to New York's Hayden Planetarium around age 12 netted him a copy of the *Skalnate-Pleso Atlas of the Heavens*, which showed the positions of the naked eye stars and then some, plus a plethora of double stars, clusters, nebulae, and galaxies. When a banged-up knee kept him off the baseball field, he spent a whole summer documenting and cataloguing the sights to be found in every constellation of the sky, and wrote them all up in a binder that he still not only has, but occasionally consults!

MAINTAINING THE FLAME

Catching the spark is one thing, maintaining the resulting flame is another. Being a kid astronomer, especially in a small town, is a lonely experience. People actually think you are a bit weird. And perhaps your parents are odd too, letting you stay up so late at night. And your classmates don't much get it either, let alone a variety of teachers. It takes the draw of the sky to offset it, the sight of Jupiter in the 'scope, the spangled path of the Milky Way, the rising of the Moon.

Once David's interest took hold, he immediately began a search both for objects in the sky and for other people who shared his interest. He sought out the Royal Astronomical Society of Canada where he encountered a number of like-minded souls. Expecting to leave with a sheaf of papers, he received only one: a *Sky & Telescope* map of the Moon, with 300 lunar craters, 26 mountain ranges, and other features. His assignment: to find on the Moon each of them and draw another map based on his own observations. That project would take the clear nights of the next two years. He finally completed his own map in 1964, and it has been published in his book *David Levy's Guide to the Night Sky.*

The next project David learned about was the Montreal Centre's Messier Club. For him, it was an adventure that would last five years, from 1962 to 1967, and involved searching for, finding, and recording notes on all of the 110 objects in Charles Messier's catalogue. This "club," started during the 1940s, was the first such observing group in North America. Then during the Summer of 1965, David was a camper at the Adirondack Science Camp about two hour's drive south of Montreal. The camp encouraged its participants to think originally, not just to follow, but to lead: to come up with bold ideas and new projects, ones that could last a long time and might actually fail.

In that time meteor observing was a mainstay of the Messier Club's activities. Our most memorable night was August 12/13, 1966, the night of the maximum of the Perseid meteors that summer. The evening was run by Isabel Williamson, one of Canada's most experienced amateur astronomers. "There was the usual overcast sky when we left Montreal," she wrote. (As one of the team remarked, we wouldn't feel comfortable if the sky were clear when we left on one of these jaunts.) We drove through the usual rain shower. We arrived at our destination and determinedly went about setting up the equipment, trying to ignore the heavy clouds. We went indoors for the usual briefing. Miss Williamson began each meteor-shower night with a careful and extensive briefing of what each observer was expected to do —and not do: "No flash pictures or you'll be shot at dawn!" She also told us about the "Order of the Hole of the Doughnut," awarded to observers who spot every hundredth meteor. At 9:45 PM EDT, one or two stars were visible and we decided to "go through the motions" for the benefit of newer members of the team. Light rain was actually falling at 10 PM when we took up our observing positions, but a few stars were still visible. The first meteor was called within the first five minutes and two more in the next five, which encouraged us to continue. Then the sky began to clear more fully. By 11:30 PM, the clouds were gone, and we enjoyed perfect observing conditions right through until dawn. In six hours of observation we recorded 906 meteors, thus breaking the group's record for all showers except the famous Giacobini-Zinner shower of 1946. It was a fantastic night.

A few months after his Adirondack experience, David had also begun a search for comets with the help of his telescope and with a book written by the comet hunter Leslie Peltier, called *Starlight Nights*. He was captivated by that book's quiet wisdom and spirituality. After 19 years of comet hunting, David finally discovered his first one in 1984. He still searches, having found his 22nd comet in October 2006. His most famous co-discovery is Comet Shoemaker-Levy 9, which collided with Jupiter in 1994, sparking the biggest explosions ever seen in the Solar System.

With no local help, Jim's experience diverged a bit. By age 12, he had a rickety three-inch reflector with which he began to find some of the treasures shown on his beloved atlas. Over the years of growing up he made friends with a variety of constellations, planets, double stars, clusters, even galaxies, and developed a passion for showing others. How best to do that than with a planetarium, something like the one he'd seen in New York. There were then none to buy, so he made one from a Crisco can that had a flashlight bulb mounted inside. There are few better ways of learning the layout of the sky than by pounding a few hundred holes in a can with a hammer and nail, guided by a hand-drawn coordinate system and with force gauged to give the right sized hole!

Jim's real astro-education began by helping at open houses at the Dudley Observatory in Albany, New York, and by joining a more-or-less local amateur astronomy club in Schenectady an hour's bus ride away, where he could discuss his passion, could hear lectures by others. A trip to the Astronomical League convention in Washington, DC in 1953 prompted him to make (with help this time) another planetarium with nearly 500 stars and a Milky Way. More than 50 years later, it resides in his basement. And it still works. The path, now clear, led Jim to an undergraduate astronomy degree, graduate school, and a professorship that allowed him not only to do research but to continue to show others, through teaching and outreach, the wonders and glories of the heavens.

AMATEURS AND PROFESSIONALS

Turn now from personal experience to the wider community. Amateur astronomy is a prime path to professionalism, to a life in which one can practice the astronomical arts all the time and get paid for it to boot. Sure, a great many professionals have trod different avenues, most others from the ethereal realm of theoretical physics. And many professionals can indeed NOT tell the difference between Jupiter and Saturn, especially when looking from beneath the canopy of a nighttime sky. One of us remembers visiting a young graduate student observing on a large telescope, who had no idea what constellation her star was part of, the other a theorist who was not sure where, or even how, to look through the telescope. But for every one of these astronomers, for whom physics is the prime attractant, there is another who not only came from the amateur ranks, but perhaps still practices amateur astronomy as well.

After a hard day reducing your Hubble data, running your latest theoretical model, computing a double-star orbit, working up a lecture, or writing a book review—or a book chapter—there is nothing quite so refreshing as strolling outdoors to admire the crescent Moon and Venus, or rolling out the 6-inch to take a look at a nebula or a group of distant galaxies. For the professional, the amateur experience brings the astronomer back to his or her roots, brings back the reason why the life was chosen, helps make the connection over the ages to all those who came before, from the time of ancient Greece through the Copernican revolution, the first discoveries by Galileo, the building of Palomar, to the Hubble and the other craft that carry the great space observatories. A few minutes outdoors under the Milky Way, admiring its dark clouds, imagining Sagittarius shooting his arrow into the Galactic heart, lets us forget the cares of the day and tells us again why we became astronomers. Amateur astronomy thus becomes far more than a path to professionalism. Once learned, it becomes a lifetime source of cosmic renewal.

But what do we even mean by "professional"? The range is both astonishing and underappreciated. The ranks run from university research professors to pro-amateurs who make livings by doing amateur astronomy, by writing, lecturing, and educating. It is then no great leap to include true amateurs who do research and teach just for the love of it alone. And that leads us to wonder what is even meant by the term "astronomer." From those who wish to admire the nightly turning of the Dippers to those who try to understand the most elegant realms of Einstein and cosmology, we all love our subject. Hence the word itself, "amateur," derived from the Latin word for "to love." Are we not then all astronomers? Are we not then all amateurs? Are there any real distinctions? Could not any of us write this chapter?

FLYING SPARKS

What brings people, from youngsters to the elderly, into the amateur fold? What else might serve as sparks? We started with two. But there are certainly as many as there are individuals. For others, it could range from a first look at the Moon through a small telescope, the viewing of an eclipse, a photo of a galaxy in a book. The list that leads to astronomical revelation is nearly endless.

The main path, though, is also the simplest. Even within the city, the sky is there for anyone to admire. And it's free. All it takes is someone to get you to look, look up in the sky. It's not a bird or an airplane up there; it's not Superman; it's the Moon or Venus or a star or even the Sun. One of two ingredients, and best perhaps both, must then be present: curiosity (what is that thing?) and/or the sense that you are seeing something unworldly and beautiful. Here is a melding of art and science, one route or the other or both, the two difficult at times to separate. Once caught within these fabrics of the Universe, you are lost, wanting only to know and experience more.

Then once discovered, the sky keeps drawing you back. You get some constellation maps, try to find the odd connect-the-dot patterns. Nothing in the heavens above seems to match. But suddenly there is the Dipper gliding nearly overhead. Aha! A signpost. Follow the Pointers, and the North Star is no longer an abstract mystery. Follow the handle to Arcturus, and re-discover that orange star. Farther down, see Spica, while rising in the northeast is Vega in Lyra, then Cygnus, then Andromeda. The flow of stars and constellations now illuminate the seasons: galloping the sky in autumn is Pegasus; in winter Orion hunts his prey, while in spring Leo rules. And we need not confine ourselves. From the southern hemisphere, the Southern Cross leads the bright stars of the Centaur across the sky in a majestic parade, while the Milky Way, its center at the zenith of heaven, is unparalleled, all such sights presenting celestial pageants that once witnessed will never fade, even over a lifetime.

Charming and beautiful in their own right, constellations also provide the sky's framework, from which you can dig deeper. Now you begin to find things that lead you from history and art into the real science. Follow the planets as they move through their starry homes, see them loop back and forth as the Earth passes by, learn to recognize them on sight. Watch a comet drifting from night to night against the ancient starry patterns. Find the naked-eye clusters, the Pleiades, Hyades, Beehive, see the Andromeda galaxy so far off in the distance, your vision incorporating billions of stars in one glance.

Now you must dig deeper, go beyond the naked eye. You get a telescope. It doesn't have to be a very good one, just enough to get into the next celestial layers. The first target is the Moon, covered with craters and lava flows, then it's on to Jupiter with its cloud belts and the same satellites that Galileo first saw, then to Saturn with its glorious rings. You look at your favorite clusters again and are overwhelmed by the sheer numbers of stars. You point your new friend to the Andromeda galaxy and find its fainter companions. Then you are ready for even further explorations. Getting detailed maps of the sky you find one thing after the other, nebulae, more clusters, more galaxies. Probing fainter and farther, you know you just have to make this a life's work: amateur, professional, a blend of the two, it makes no difference.

At some point along the path, though, it's hard to go it alone. It takes the support of others. And where best to get it other than within the wider astronomical community, from other amateurs and astronomy clubs where you can meet lots of like-minded souls or even form your own group. Now you can share, your experiences for theirs, observe together, get books about the things you see, and teach each other. Excited by what they learn and know, club members (including you) reach outward, to schools and teachers, and send their own sparks back into the community at large, from which the next generation of astronomers and scientists will come.

The club brings in speakers, professionals, and if beginning your career path, you begin to see how the pro-astro life might really work for you. On now to college to study the stars, on to grad school, on to employment where you glory in being an amateur astronomer full time, not just at night but during the day too, as you continue the research you began as a youth and as you teach yet others, some of whom will catch the spark from you. And so you pass it down, the one from Grandma or from your first meteor, the cycle repeating much as does the Galaxy's cycle of life from star birth to death and back to life again. And all because someone got you to look up at an orange star, which is now but a metaphor for all of our experiences in discovery.

In the wider environment, our society needs, thrives on, science. Not just astronomy, but science in general. Without it, and the new discoveries it brings—engineering, new products, better lives for us all—society stagnates. Where do the new scientists come from? Many, if not most, rise from the experience of childhood curiosity, from the sparks that turn them into first amateurs, then perhaps for some, into professionals. The sparks are many: bugs, dinosaurs, the stars. No matter what captures the young mind, the world of science suddenly opens for exploration. Discovery of the fascination of the heavens can then lead to exploration of other fields of fascination, the heavens pointing the way to biology, medicine, chemistry, all the areas of science that society needs in order to flower and grow.

Only a few who catch the scientific fever will go on to become the professionals. Most will go into other work: business, law, education. But the spark will have lit the fire. Here are the ranks of amateurs, those who love the subject, from casual outdoor aficionados to dedicated telescopic observers. Here is the true core of science, those who can extend the gift of the spark, who provide the financial support, who pay the necessary taxes, and who write their congressional reps to make sure that Hubble survives. We cannot do without them; science, the love of science, depends vitally on them.

So we pass down the sparks from one person to another, the sky perhaps creating the most powerful of them. Amateurs thus provide the foundation on which we build, in teaching, in sparking, in mentoring, in appreciating the glories of what nature gave us. And all because someone got them to look up. Up in the sky. Perhaps at a meteor. Or at an orange star.

2

Sowing the Seeds of Science: Amateur Astronomers Engaged in Education and Public Outreach

Scott Roberts

The Milky Way galaxy with its billions of stars, our own mother Sun and its offspring of planets is our galactic home and as part of it we are moving at incredible speed through space on a fantastic cosmic journey amidst a universe filled by countless other galaxies. And while even Monty Python in their Galaxy song in the movie *The Meaning of Life* tried their best to contribute astronomy to the public understanding of science, the distances, size, age, and complexity of celestial bodies are difficult to comprehend for almost everybody, and even though most people today have some knowledge of the universe, few feel that things so far away as the stars and galaxies have any real significance in our daily lives.

But the history of humanity reveals that our understanding of the stars has played a fundamental and critical role in our lives that extends to the present day. Throughout the ages we have been unraveling many scientific mysteries in our journey of astronomical exploration and discovery. In virtually every culture around the world we find that studying the stars was of the utmost importance. A few indications are petroglyphs created by early humans in North America that indicated the solstices; the 5,000-year-old Stonehenge monument near Amesbury, England used to determine changes of the seasons; entire cities built, for instance by the Maya more than 1,000 years ago in accordance to astronomical alignments to Earth's orbit around the sun; the sophisticated 2,000 year old computer called the Antikythera Mechanism used to calculate motions of stars and planets; and meticulous astronomical records maintained by the Chinese dating as far back as the sixth century B.C. Mapping the heavens and creating constellations from the patterns of stars in the sky began over 4,000 years ago. As we studied the cosmos, our cultures, mythologies, religions, art, laws, philosophies, and scientific knowledge became intertwined, forging the societies of modern humanity.

Our ability to understand the changes of the seasons and feed ourselves, tell time, explore uncharted regions, or govern society can be traced to what we have learned from our efforts to study the stars. The prosperity and survival of our "modern" societies is so dependent upon our knowledge of the heavens that without it we may well still live the lives of hunter gatherers. Professional and amateur astronomers understand how critical it is that we continue to explore the cosmos as part of the human quest for exploration, discovery and our thirst for ultimate explanations. Technical advances combined with an historically unprecedented number of professional and amateur astronomers are leading to such a rapid rate of discoveries that one could argue that we are in the midst of a golden age of astronomy. But in order to continue its work and gain greater insight, the astronomical community needs the support of society for continued funding of education in science, astronomy, and space exploration, as well as major scientific projects.

Yet recent public surveys commissioned by the U.S. National Science Foundation and Eurostat, the European Union's equivalent of the Census Bureau indicate a general decline in interest in, and attentiveness to science with the general public, particularly in the United States and in Europe. This is causing real concern among high-tech and pharmaceutical companies as they look at their potential to develop with better technology and advances in medicine with the next generation as evident from the stir of articles over the last few years of the long-term impact of the dearth of science in our schools. The growing lack of interest in science could be somewhat attributed to our own progress. According to Globe@Night, a citizen science project that measures light pollution, in our light-polluted communities, fewer than one in four people have an opportunity to ever see the glow of the Milky Way stretching across a dark, starry night sky. Many are not even aware that the Sun is a star or that its energy is the fuel of life, and according to biannual surveys conducted by Jon Miller and colleagues for the National Science Board, just half of the public are aware of the fact that Earth orbits the Sun rather than the other way around. Sadly, most of society today is so preoccupied with their busy daily lives that many have disassociated themselves from nature, arguably at the expense of our own collective good; so much indeed that a new term, nature deficit disorder, has been coined in a recent book by Richard Louv entitled *Last Child in the Woods: Saving Our Children from Nature Deficit Disorder*. Clearly, many have lost touch with one of the most basic and yet most fascinating activities of science: being part of and observing nature.

It was Carl Sagan who said, "Earth is a very small stage in a vast cosmic arena." Just looking up at a star-filled sky humbles most of us, opens our minds, and with just a little astronomical background knowledge confronts us with a profound truth that we are somehow interconnected with the cosmos. Those who encourage others to join in on the adventure of exploring the sky and the universe are engaged in astronomy education and public outreach. While there are many who make education and public outreach their profession in classrooms, science centers and museums, public observatories, and other institutions devoted to astronomy, the majority of this work falls to thousands of amateur astronomers, who do it without compensation of any monetary kind. Many are self-educated and trained, and they own portable telescopes that they use to share the adventure and their love of astronomy with others. Amateur astronomers are often eager to share the experience and will patiently guide and teach people they have never met before how to use a telescope so that they can explore and discover the beautiful celestial treasures of the cosmos for themselves.

While sharing the eyepiece of a telescope to show others the belts of Jupiter, or the faint glow of a distant galaxy is fun and interesting, exposing others to astronomy has many other direct benefits. The very act of contemplating the vastness of the universe calms people down, and creates some space within to put priorities into perspective. It can motivate young people to pursue careers in technology, medicine, engineering, and mathematics. And it can lead to increased interest, attentiveness, and even understanding of astronomy. Ultimately astronomy can motivate people around the world to support bold programs of exploration and discovery.

Getting started in astronomy outreach can be simple. All one has to do is set up a telescope around other people, point it to the stars, and invite them to look. Most who try find that they instantly take on the role as interpreter and guide, and their telescopes become a vehicle, transporting newfound explorers across the cosmos. A flood of questions can arise and a learning process may begin. Of course, as Chapter 4 describes, not every observing amateur astronomer feels comfortable engaging with an inquisitory public, though, as Chapter 5 shows, most are sufficiently knowledgeable to satisfy their audiences' thirst for astronomy knowledge.

If you ask people what they felt when they looked up at a night sky you'll find that the experience has touched them somewhere deep in their minds and hearts. They will often ask how did it all begin, how did we get here, what is happening now, and where are we going? Once questions like these arise, an age old journey of exploration, discovery, and learning has started, and much anecdotal evidence and few systematic studies tell us that amateur astronomers can make excellent guides and interpreters, as Chapter 6 explores.

The more someone masters their skills with the telescope and their interpretation of the sky, the more they can educate and share the experience through outreach. Showing someone the rings of Saturn or a distant galaxy for the first time brings about a deep satisfaction that is difficult to express in words. Outreach amateurs can remove others from their daily existence and transport them across time and space as they explore other worlds. And their realization that we are on a tiny planet orbiting a small star in a galaxy containing billions of other stars often strikes them to the core. Many amateur outreach enthusiasts have witnessed people who are looking through a telescope at another planet become wide-eyed and open-jawed. Understanding astronomy awakens people to the fact that there is an unbroken connection between themselves and the furthest reaches of the cosmos, between the past and a future far beyond the existence of our home planet.

What we have learned from amateur astronomers is that one doesn't have to be a research scientist to make discoveries, or an astronaut to become an explorer, nor does teaching others to do the same require a degree. Sharing what one knows and feels passionate about with others comes quite naturally to many and the following chapters explores this in great detail. But there is more to education and public outreach than sharing a passion with others; in the process of educating others, the transformation towards becoming a teacher has begun.

Outreach in astronomy seems, on first sight, to be a selfless act, but for those who choose to embark on this path, it may quickly become apparent that personal growth and personal satisfaction make outreach something of a selfish activity. There are few things more gratifying than opening the minds of others, fostering a sense of wonder, providing greater understanding, and setting them on a never ending journey of learning. Amateurs involved in outreach are helping people from all walks of life to understand the importance of astronomy and we hope that more will join in this quest.

3

Working with Schools: Teaching Students on Their Own Turf

Tim Slater

There is a long and rich tradition among amateur astronomers conducting outreach by "reaching out" to the public by purposely setting up telescopes on busy public sidewalks, in parking lots outside of grocery stores, within National Parks campgrounds, and near the snack bar at a county fair. In these informal learning settings, amateur astronomers often easily attract the curiosity of passersby with a mere smile and a wave, accompanied by a promise of glimpsing something wonderful through a mysterious contraption. These informal invitations to look through a telescope are scattered widely and without expectations among either the issuers or the receivers and represent a win-win personal *Encounter of the Third Kind*.

However, beyond this kindly and informal environment of the roaming mini-star party, there lies a realm that only some amateurs dare to enter—a realm of uncomfortable desks in repeated straight rows, where graded performance is necessarily emphasized more than harmless curiosity, where everyone's behaviors are governed by strict rules with consequences, and where attendance is legally mandated. Despite the natural uncomfortableness of entering a learning environment where the participants might not wish to actually participate, schools represent an important location where an amateur astronomer can interact with tens, if not hundreds, of students. School venues can be overwhelmingly successful when amateurs who work with schools understand the important and sometimes surprisingly subtle aspects of school culture. This chapter provides a brief overview of where school-based formal education fits in the spectrum of education and outreach and describes critical need-to-know information that most readily helps amateur astronomers be successful partners with school children, parents, teachers, and school administrators.

THREE REALMS OF EDUCATION AND OUTREACH

In the vernacular of education and outreach, we roughly, and somewhat artificially, group the different ways that amateur, and professional, astronomers bring their enthusiasm and knowledge of astronomy to the masses into three broad and overlapping categories. These learning venues are most often described as (i) formal education, (ii) informal education, and (iii) public outreach. Although amateur astronomy itself seems to defiantly resist categorization, it probably most commonly sits comfortably in the seam between informal education and public outreach.

Considering the domains shown in Figure 1 somewhat unconventionally from right to left, public outreach is often characterized as activities where

Formal Education	Informal Education	Public Outreach
Participants are required to attend	Participants choose to attend	Information is delivered to participants
Fewest number impacted	←——————→	Greatest number impacted
Interactions are longer in duration	←——————→	Interactions are shorter in duration

Fig. 1. Three Overlapping Realms of Education and Outreach
(Adapted, with permission, from Cherilynn Morrow, 2003.)

scientific information is delivered to the public *in situ*—that is, people's private homes through cable TV, in their cars over the radio, and in their doctors' waiting rooms through magazine subscriptions. Public outreach is delivered in relatively short bursts to incredibly large audiences in one-size-fits-all approaches, with little or no direct interaction with the receiver. In this sense, public outreach delivered via video can take the form of 40- to 60-minute video documentaries and IMAX films, such as those found on the *Discovery Channel* and the *Science Channel*. It can also be found on the radio in the form of two-minute audio shorts, such as those known as McDonald Observatory's *STARDATE* or *Earth and Sky*. Of course, newspapers, news magazines, and astronomy-specific magazines share the excitement of astronomy on a regular basis. Even the quickly growing and evolving Internet participates in public outreach too, particularly in the form of websites such as *Space.com* and *UniverseToday.com* or through Internet-delivered Podcasts such as *Astronomy Cast* or *Slacker Astronomy*. All of these examples, and many others, can be most easily found by conducting an Internet-based search using a common search engine, an example of which is *Google.com*. For amateur astronomers, these public outreach activities typically provide important resources for amateur astronomers to use, rather than represent things that amateur astronomers actually do themselves.

Whereas public outreach is generally considered to use the approach of "take it to the people," the approach of informal education relies on "build it and they will come." Informal education is generally considered to be the domain of museums, planetariums, nature centers, visitor centers, and comprehensive science centers. These informal education facilities and/or programs vary considerably from one to another in how many people they serve as well as how broad their subject areas are. As a few examples, New York's American Museum of Natural History covers nearly all of the science disciplines whereas Chicago's Adler Planetarium and Astronomy Museum or the Smithsonian Air and Space Museum focus on the much narrower domain of exploring the universe of topics beyond Earth's atmosphere. For amateur astronomers, these informal education facilities can serve as important rallying points to host monthly club meetings or to utilize their mailing lists to advertise star parties designed to share the wonder of the night sky. Amateur astronomers often play a critical functional role for museums working as docents, or volunteer tour guides, to enhance the experience of visitors.

What most easily distinguishes public outreach from informal education is that informal education relies heavily on the concept of *free-choice learning*. Free-choice learning was widely popularized by John Falk, who wrote a book of the same title. Free-choice learning is the type of learning guided by an individual's needs and interests and is an essential component of life-long learning. Indeed, it is likely that free-choice learning is the most common type of learning that people engage in. In fact, free-choice learning is so common that it is most often simply taken for granted. Because people have control over what and how they learn, and because they can choose to learn in appropriate and supportive contexts, free-choice learning can be highly efficient and the ideas acquired this way can have long duration.

It is in the informal education venues of parks, parking lots, and planetariums that most amateur astronomers share their unique outreach talents with a public who chooses to attend and participate.

Many also step through the doors of formal education institutions, an environment that for some may be unfamiliar and intimidating in these days of standardized testing, contemporary science teaching methods, and increased safety concerns.

FORMAL EDUCATION

The world of school buildings and classrooms characterize the classical domain of formal education. For most of the United States, children are required by law to have 8 to 10 years of formal education learning (a significant fraction of children in the United States experience "home schooling," a discussion of which is far beyond the scope of this short chapter; however, home schooled children can often be found participating in informal education venues and programs). In other words, schools represent the part of the education and outreach spectrum where students are actually forced to endure learning about science. As such, there are some important barriers that need to be recognized and overcome for amateur astronomers to successfully and positively help schools improve their science education.

What is Taught?

What surprises many amateur astronomers who become engaged with schools is that the long list of science topics to be taught in schools across grade levels will vary considerably from state-to-state, district-to-district, and sometimes even classroom-to-classroom! Although there are documents floating around schools referred to as National-standards or State-standards, there is actually no nationally mandated curriculum or list of topics that all students must learn in the United States. To make a great generalization across the country, what is typically found in most school written guidelines is that students in elementary grades kindergarten through 4th grade (roughly ages 5 to 9 years) learn about objects in the sky as well as how and why they appear to move. In the middle grades of 5th grade through 8th grade (roughly ages 10 to 13 years), students are learning about objects in the solar system and the movement of objects around the Sun. Finally, in the high school grades of 9th through 12th (roughly ages 14 to 18 years), the more conceptually abstract ideas of astronomy including stellar evolution and the origin and the structure of the universe including the Big Bang are included.

Fortunately, or unfortunately depending on your point of view, these guidelines are sometimes faithfully followed and other times rarely followed. This happens for a whole host of perfectly good reasons, but can be most often attributed to the fact that only a small fraction of students in the United States elect to take science coursework in high school beyond Biology, to the fact that few teachers have had much formal college coursework in astronomy and therefore lack considerable knowledge about astronomy, and, perhaps most importantly, that teachers often feel that the long list of topics that they are required to cover is too long to devote considerable time to astronomical topics. This is where the amateur astronomer can step in to lend a hand.

However, this situation is both a great advantage and a great barrier to the amateur astronomer who wants to become involved with and help schools. It is a great advantage in that many teachers have considerable flexibility in what they teach and over what duration a topic is covered—such that astronomy and its accompanying technology can play an important role in a classroom. At the same time, the documents that guide teachers, and the accompanying tests, may hamstring a teacher or a school into feeling that there is little room for anything related to astronomy and technology and any pursuit of such is a distraction and a disservice to the children's success. Indeed, the most important thing an amateur astronomer who wants to get involved with schools needs to delicately evaluate is where there is an opportunity to bring astronomy into the classroom without causing teachers to feel like their students might be disadvantaged on the state tests.

What Does Classroom Instruction Look Like?

Quite different from the days gone by of when the average-aged amateur astronomer was a student in school, the nature of what classroom instruction looks like today has changed. Certainly more-so in some schools and less-so in others, teachers in the 21st century do not often lecture for hours on end and ask students to take copious notes for memorization. Instead, there exists a somewhat different perspective of how best to help students learn. Amateur astronomers who wish to be successful in schools need to understand and utilize this perspective.

Contemporary science teaching is based upon a philosophical foundation known as *constructivism.* In grossly abridged terms, classroom instruction built upon a constructivist viewpoint leans heavily on the idea that students already think they know a considerable amount about how the world works—they believe they understand why the Moon appears to change shape over the month (the Earth's shadow blocks the Sun's illuminating light), why it is hotter in the summer time than the winter time (the Earth's closer to the Sun in the summer), and that gravity is caused by a planet's atmosphere (the Moon has no atmosphere therefore cannot have any gravity), just to name a few of the most tenacious misconceptions. In other words, the teaching perspective is one that could be paraphrased as, "determine what each student knows and teach them accordingly." This is in direct contrast to the more traditional viewpoint of yester-year that students don't really know anything, they are essentially blank slates upon which knowledge is to be precisely written, and it is the teacher's responsibility to clearly explain everything to them so they will then know the things we most want them to know.

This has important ramifications for the amateur astronomer teaching in today's classrooms if they want to be effective. In days gone by, when a guest speaker came to visit a classroom, the students would most often respectfully listen to the speaker, no matter what was being said or how it was being said. However, students of today fully expect, if not demand, to understand *why* what they are being told has value for them in order for students to stay tuned-in to the presentation. The most important aspect of contemporary classroom-based instruction is that it is CRITICAL for classroom visitors to directly and intentionally ask students what they think about an idea and what pre-existing knowledge they bring to the class about the topic before "teaching" or else students will quickly become bored and lose their attention after the visitor's novelty has worn off. The key piece here for the effective amateur astronomer teaching in a school is to ask students a lot of questions and give only a few answers.

WHAT CAN AMATEUR ASTRONOMERS UNIQUELY TEACH?

First and foremost, amateur astronomers are uniquely positioned to bring an unbridled enthusiasm for observing the astronomical world to students. All too often, students in formal learning environments spend their intellectual energy on "just the facts, Ma'am" and sometimes miss the excitement of discovery and the pleasure of observing. Personal stories of why amateur astronomers do what they do make science appear to be more human and less abstract. Students need to be exposed to why someone would value staying up all night looking for small, fuzzy, Messier objects.

Second, amateur astronomers can emphasize that astronomy isn't only done by highly paid scientists at the university. School children all too often have the notion that science is only done by the best and the brightest white males and many students do not see themselves as being able to be an important part of the scientific enterprise. Amateur astronomers come from all backgrounds, all ethnicities, all professions, all socio-economic groups, and all aptitude levels and amateur astronomers are uniquely situated to carry this message to students—the message that science has something for everyone!

Of course, which of the more traditional academic learning concepts can be targeted depends somewhat on the venue. In formal, day-time, classroom settings, a slide-show tour comparing and contrasting planets or planetary nebulae works much better than telescope observing (aside from the fairly obvious solar projections showing sunspots, of course). The important piece of this is that students need to understand how astronomers know what is out there; this is many times more important to share than a long, fact-by-fact listing of how big and how far away the many objects of our Universe are. In short, a long presentation of facts will be quickly forgotten, but a discussion where students are asked to compare, for example, projected images of galaxies have considerably more staying power.

On the other hand, the learning concepts that can be targeted during a nighttime star party are fundamentally different. Although a computer-video projector showing images from a telescope-mounted CCD camera are becoming an evermore common curiosity at star parties, the chance for students to look through a telescope offers unique learning opportunities that do not happen in conventional sit-at-the-desk classrooms. The primary educational goal of a star party is one of gaining experience. Somehow, people seem to readily recall the first time they looked through a telescope (my personal recollection is of seeing the rings of Saturn). After gaining experience, the secondary goals are more often about the technology and optics involved with a telescope rather than learning facts and figures about the objects being observed. The common theme across all of this is that the process of doing astronomy is significantly more educationally enduring than remembering numerical data.

OBSTACLES TO SUCCESSFULLY WORKING IN SCHOOLS

It might seem incredible that a knowledgeable amateur astronomer who enthusiastically volunteers to help a school with their science education efforts might be politely told, "No, thank you." It is most certainly not that schools do not need nor desire the help of amateur astronomers. Rather, there has been a long history of science-minded folks who have walked into classrooms to "help" without understanding the important aspects of school culture they were trying to help with. In this section, we'll explore a few of these with the goal of increasing awareness.

No Child Left Behind

For the past 200 years, schools in the United States have been coordinated, directed, and run by local school boards with only some state-level controls and essentially no national-level federal controls. In the past decade, this has changed dramatically with unprecedented federal-level legislation known as No Child Left Behind (NCLB). NCLB requires that all schools demonstrate improvement in student achievement each and every year through standardized-testing and that every student is required to be taught by a highly qualified teacher. These are important goals that nearly everyone would agree are important in principle, but have far reaching implications when put into practice. For the amateur astronomer, the most important aspect of NCLB is to recognize that today's students are tested, tested, and tested and that these test scores really do matter. As such, teachers and schools feel considerable pressure to focus on helping students increase their achievement test scores above all else—even if that "else" includes unique opportunities to learn about astronomy and technology. In other words, teachers are pressured to teach what material is covered by the mandated achievement tests. The enthusiastic amateur astronomer needs to determine the extent to which students learning about astronomy will help improve achievement test scores before delving too far into a particular school's environment because this varies considerably school-to-school and state-to-state.

Certification and FBI Clearance

Some school districts require that anyone who is working with students hold formal, state-approved certification or licensure to do so in any sort of formal way. Even more common, some school districts require people involved in classroom instruction to have received an FBI Clearance (commonly known as a *Fingerprint Card*) certifying that they are not criminals with felony-level records or are a registered sex-offender. Amateurs working with schools need to contact the school principal's office to determine if they need to have some sort of certification or FBI clearance in order to provide instruction to school children. These items are not normally a barrier, but do pop up unexpectedly more often than some amateur astronomers might expect.

Inclusion of ESL and Special Needs Children

The diversity of students found in science classrooms today has never been wider, deeper, or more exciting. For example, in many regions of the country, teachers have the responsibility to teach students whose primary spoken language is not English. These students are sometimes referred to as ESL-students: English as a Second Language. In a similar way, some amateur astronomers are caught by surprise when they encounter students who have special needs, including those that are visually impaired, hearing impaired, or otherwise physically handicapped. Sometimes these students have a school staff member assigned to them to help, such as a sign-language interpreter, and others can have high-tech assistive devices. Regardless, schools have a legal and moral responsibility to educate all students. An amateur astronomer has to think carefully about how to describe an astronomical object to a student who can't hear you speak except through an interpreter or how to show a beautiful picture of a galaxy to a student who can't see it. It is appropriate to ask a teacher if there are students in the class who have special needs and ask the teacher for specific suggestions on how to help these students experience the excitement of astronomy too.

NUTS AND BOLTS FOR SUCCESSFULLY
WORKING IN SCHOOLS

Teachers and schools benefit greatly from engaging members of the community directly with their students. Unfortunately, this is most often limited to a local "career day" where doctors, electricians, pilots, store managers, cosmologists, and cosmetologists allocate a day to tell students about their daily work lives. Imagine the excitement that could be generated in students by people sharing their unpaid passions of bird watching, fossil digging, and, of course, astronomy! Working with amateur astronomers provides numerous benefits to teachers. At the top of the list, teachers can greatly benefit their students by having amateur astronomers demonstrate what life-long-learning, in contrast to school-book-learning, really looks like. Further, teachers will appreciate having an individual with expertise that they can call on when students have questions that need a prompt reply. Additionally, school budgets rarely allow for equipment purchases like telescopes; the physical resources amateur astronomers can bring to the classroom are highly valued. Alternatively, schools sometimes have had telescopes donated to them in years past, which desperately need some basic repairs and fine tuning in order to be used. In short, the benefits to teachers for having amateur astronomers in the classrooms are immeasurable; however, the cultural gulf between amateur astronomers and classroom teachers needs to be carefully navigated and successful relationships require some understanding of the nature of teachers and schools.

ASP's Project ASTRO program, described more fully in Chapter 6, capitalizes on the respective strengths of teachers and astronomers by training them to work successfully as a team.

Communicating with Teachers

Although most teachers have an email address and many have telephones in their classrooms, teachers typically have their hands completely full during the school day working with students. As a result, most amateur astronomers find it is easiest to contact teachers at home in the evenings or to realize that emails might not get answered as quickly as one might normally expect. It is recommended to contact teachers as far in advance as possible and to be understanding when communication takes longer than one hopes.

Security at the School Site

Many schools have very strict protocols about who can be on the school grounds and how people gain access to classrooms. In general, a visitor should ask their partnering teacher where to park and how to get a visitor's pass. The most common strategy to get a visitor's pass is stopping by the Principal's office to show picture-identification and to sign in and out in a visitor's log book. Some schools will not let anyone walk around without an escort. Similarly, some schools have metal detectors. In other words, be sure to arrive at the school plenty early to gain access.

Evening Outings

Amateur astronomers have found varying success with planning evening star parties at school sites. On one hand, some have found that regardless of how interesting the program sounds, the weight of other activities, such as soccer practice, makes it so that not a single additional event can be squeezed into students' lives. On the other hand, an evening of star gazing is innately appealing to some students and holding a star party at a school can be an amazingly popular event. One thing to remember is that few schools seem to have the ability to turn off many of the outside, night-time security lighting, so the site needs to be surveyed before any evening event can be fully planned.

Capitalizing on Previously Scheduled School Events

One way to get around planning a star party, which might be viewed by parents as an imposition to attend, is to coordinate and hold a telescope viewing during a night where there is already a school-based event going on, such as a basketball game, a science-fair, or a school open-house.

Similar to setting up telescopes at the county fair near the snack bar, a telescope set up outside near the school's entrance is certain to get a lot of traffic (and might even be more interesting than the program going on inside!). The school principal's senior administrative assistant will have all the information and a school calendar and can help plan which event would be the most appropriate.

Capitalizing on Celestial Opportunities

There is a long tradition in astronomy education and outreach to exploit rare events in the sky to attract attention of the public. These events serve as an excuse to bring school children out to look through a telescope and up at the night sky. Examples of successfully leveraged observational opportunities include solar and lunar eclipses, meteor showers, close planetary oppositions, comets, star cluster occultations by the Moon, and space probe encounters.

EPILOGUE

Without question, amateur astronomers are uniquely well-poised to bring an enthusiasm for science to school children. Although star parties hosted in locations such as busy grocery-store parking lots or at the local county fair have been the meat-and-potatoes of amateur astronomers for sharing the night sky with interested parties, there is a unique opportunity to share astronomy by partnering with schools. Indeed, some important programs have successfully rallied amateur astronomers to partner with school teachers (as mentioned earlier, ASP's Project ASTRO is a great example). However, one should not underestimate the tremendous power a citizen scientist has in igniting a spark within a child to become more curious about the natural world in which we live.

REFERENCE

Bybee, R.W., and Morrow, C.A. (1998, Fall). Improving science education: The role of scientists. Newsletter Forum Education American Physics Society.

4

Attributes and Practices of Amateur Astronomers Who Engage in Education and Public Outreach

Martin Storksdieck and Marni Berendsen

WHO ARE AMATEUR ASTRONOMERS?

By day they work as accountants, mechanics, physicians, clerks, software engineers, or teachers, fitting in seamlessly into the population, unnoticed by their colleagues and peers, but by night they shed their day jobs and become star gazers, comet hunters, telescope makers, and club enthusiasts, passionate about the night sky and what lies beyond the flimsy atmospheric boundary that creates a fragile island of life on our home planet Earth.

We really don't know for sure how many amateur astronomers (or those who would call themselves such) are currently active in the United States, and the number depends on who is included. In general, amateur astronomers are hobbyists who pursue an astronomy-related passion as part of what the Canadian sociologist Robert Stebbins terms *serious leisure* (Stebbins, 2006). The Astronomical Society of the Pacific identified three somewhat overlapping categories of amateur astronomers:

Research-level amateurs are *citizen scientists,* non-paid experts that use their often sophisticated telescopes and related technology, including their computer power, to support professional astronomers in research programs that mostly require data collection, processing, or analysis by a large number of individuals. There are probably only a few hundred research-level amateur astronomers who work intimately with professional astronomers or on their own (Fraknoi, 1998). Far more, in fact potentially thousands may contribute to broader citizen science projects such as Globe at Night, Seti@home, or Stardust@home.

The second group are **astronomy enthusiasts**, a group of amateur astronomers who pursue their interest in astronomy by reading astronomy-related articles in newspapers and magazines such as *Sky & Telescope*, *Astronomy* or *Mercury*, buy and read popular books on astronomy, attend public lectures on astronomy-related topics, watch television programs related to astronomy, or visit IMAX theatres when astronomy and space programs are featured. Increasingly, this group may follow astronomy-related topics and issues on the internet, where a plethora of high-quality websites, blogs, and listservs on astronomy provides ample opportunity to pursue one's interest in astronomy from home. This group may include many who do occasional observation of the night sky, either with telescopes and binoculars, or with the naked eye. They may know constellations and major objects in the night sky, but observing the night sky is not their main passion. Like research-level amateurs, they may belong to an amateur club, but club membership may not be the focus of many of their hobbyist pursuits. Astronomy enthusiasts are the largest group of amateur astronomers, though even a rough estimate of their numbers is difficult given the rather fuzzy definition on who may belong to this group. The combined unduplicated circulation of the two major astronomy magazines (*Sky & Telescope* and *Astronomy*) suggests that there are about 300,000 to 400,000 enthusiasts (Feinberg, 2006), but the total number of people who are highly attentive to astronomy may be considerably higher (NSB, 1998; 2000; 2002).

The third group of amateur astronomers can be classified as **observing amateurs**. This group will resemble the astronomy enthusiasts in many ways, but they differ in one crucial aspect: they tend to own a telescope (or many) and regularly use their telescope(s) to observe the night sky or the sun. They may do that on their own, share their passion with family, friends, or the public at large (through events like public star parties), and they tend to be more likely to join amateur astronomy clubs and associations.

Many observing astronomers are members of astronomy clubs (though, admittedly, it is unknown what the share is who go it alone). There are about 750 amateur astronomy clubs in the U.S. with a total membership of about 50,000 (ASP, 2005).

CHARACTERISTICS AND INTERESTS OF AMATEUR ASTRONOMERS

Most of what we know about amateur astronomers and their outreach stems from a 2002 web-based survey, funded by the National Science Foundation and conducted by the Institute for Learning Innovation and the Astronomical Society of the Pacific (Storksdieck, Dierking, Wadman & Cohen Jones, 2002). The data reported in the following section are mostly based on this survey of more than 1,100 amateur astronomers.

The vast majority (83%) of amateur astronomers who responded to this survey were male and more than half were between 31 and 50 years of age. Another 30% were 51 to 65 years of age. Even though we cannot assess with certainty how representative this survey is, the results still suggest that the typical amateur astronomer may be a man between the age of 31 and 65.

Amateur astronomers seem to have a higher level of education than the general public, particularly in terms of astronomy and science. More than two-fifths of the amateur astronomers indicated they had some formal training in astronomy, astrophysics, physics, or other related science field, and two-thirds had some form of informal training in those areas. In fact, three-quarters of those responding had either formal and/or informal training specifically in astronomy-related subjects.

The majority of amateur astronomers who responded to the survey have been interested in the subject of astronomy for more than 20 years. Relatively few amateur astronomers seem to be new to the field of astronomy. And these amateur astronomers reported a wide range of interest in matters related to astronomy, though traditional observation of objects in the night sky was the most frequently mentioned specific interest (90% of the amateur astronomers responding indicated this choice). Almost four-fifths expressed high interest in general astronomy and current events in astronomy. Interestingly, amateur astronomers expressed interest in public education, teaching and sidewalk astronomy at a similar level (65% highly interested) as their interest in the solar system; stellar evolution, novas, supernovas, etc.; astronomical instruments like telescopes, CCD cameras; or telescope making. Slightly lower ranked in the overall interests of amateur astronomers were environmental topics like light pollution and the greenhouse effect, history topics related to astronomy, and folklore, mythology, and story-telling.

AMATEUR ASTRONOMERS AND EDUCATION AND PUBLIC OUTREACH: WHAT, HOW, AND WITH WHOM DO AMATEUR ASTRONOMERS SHARE THEIR PASSIONS

The 2002 web survey of amateur astronomers suggests that a significant proportion of amateur astronomers, particularly the ones organized in clubs, pursue their hobby by sharing their interest and enthusiasm for astronomy with others in their communities. Education and public outreach (EPO for short in the lingo of the informal astronomy education community) involves a diverse set of activities, including (but not limited to):

- scheduled star parties (where the public is invited to a park or other venue for an astronomy program with telescopes);
- sidewalk astronomy (where the amateur astronomer sets up on a public sidewalk and passers-by have an opportunity to peek through a telescope);
- classroom visits;
- presentations and activities at events for Boy or Girl Scouts or other organized youth groups, as well as for a variety of community organizations;

Amateur astronomers may also act as volunteer interpreters and educators at a local science center, museum, planetarium, library, church, or science fair. Public outreach events also center around well-publicized astronomical events such as eclipses, comets, or meteor showers. Events occur not only in low-light and rural areas but also in highly urban areas, where planet and lunar viewing are especially well-suited to light-polluted environments.

The most commonly reported outreach events are public and school-based star parties and presentations in classrooms, but amateur astrono-mers might reach a large public by writing articles for local and regional newspapers and magazines, appearing on local radio and TV programs, organizing an Astronomy Day or an Earth Day or featuring their craft and passion at county fairs. Amateur astronomers conduct a consider-able amount of EPO as "inreach," or education and learning that happens within the club, when members share their knowledge and skills with other club members (Storksdieck, 2005).

The following table (from Storksdieck et. al., 2002) presents a break-down of outreach activities reported by amateur astronomers.

Table 1. *Type of Educational Outreach Activities*

Type of Educational Outreach Activities	Percentage (n=642)
Star parties	82%
School-related activities (formal education)	81%
Presentations/workshops	58%
Conduct or organize internal training for club members	35%
Volunteer work for "museums" etc.	29%
Other	19%

Note: Multiple answers possible; totals exceed 100%.

Based on membership and activity level in the Night Sky Network (Storksdieck, 2005), the 2002 comprehensive outreach survey conducted jointly by the ASP and ILI (Storksdieck et. al., 2002), and anecdotal reports from astronomy club members compiled at the ASP, it is estimated that ap-proximately 20% of current club members are at some level involved in edu-cation and public outreach, for a total of about 10,000 outreach amateur as-tronomers. The vast majority of these club outreachers (about 6,500) conduct or participate in just a few outreach events per year, up to one per month. Another 2,000 engage in activities between two and three times per month, and the remaining 1,500 are active almost every week or even more often.

Club members are not the only amateur astronomers who are engaged in EPO; a large number of amateur astronomers who are not affiliated with clubs might engage in EPO as well. Unfortunately, we are currently missing reliable estimates on their numbers and activities. However, the 2002 survey revealed that 69% of respondents who conducted outreach were also members of a club; in fact 37% fulfilled some official club function. The 2002 survey could not establish sample and selection bias, but it provides an indication that EPO by non-club affiliated amateur astronomers will not likely be vastly more than that of club-affiliated ones, and probably will be considerably less.

As will be reported more fully in Chapter 5, a survey by Berendsen (2005) found further indication that not only are unaffiliated amateurs less likely to do outreach, but members of astronomy clubs are increasingly likely to participate in outreach the longer they remain members of a club. Additional results from Berendsen's survey that have not previously been reported indicate that neither gender nor amount of formal astronomy education affect an amateur astronomer's decision to participate in education and public outreach. Instead, they are more strongly influenced by their years of club membership. Perhaps their confidence as well as their knowledge increases the longer they remain a club member.

PARTNERSHIPS: AMATEUR ASTRONOMERS WORK WITH OTHERS IN THEIR OUTREACH EFFORTS

Outreach amateur astronomers seem to partner with organizations that provide access to various target audiences. Almost three-quarters of amateur astronomers doing outreach report working with a local planetarium, observatory, and/or a local science, technology, or nature center, or natural history museum. More than half of them seem to form an informal or formal association with a school (mostly elementary or middle school), college, or school board or district. Almost a quarter were partnering with a local Recreation or Parks department.

Amateur astronomers also tend to do outreach together as a club activity, a major reason why club membership seems to encourage outreach. Only about a third of amateur astronomers engage in educational outreach individually, a quarter engage primarily as part of a group, and the rest regularly engage in educational outreach activities sometimes as individuals, and at other times as part of a group.

AUDIENCES OF AMATEUR ASTRONOMERS

Amateur astronomers serve a broad community, offering their educational activities wherever, whenever, and to whomever expresses an interest or need. Almost all (90%) stated in the 2002 survey that they serve the general public (including community groups, families with children, teenagers, and adults) and more than two-thirds work with school groups. Internal training of new club members and/or interested individuals who visit club meetings was also identified as an important audience for educational outreach.

The great majority of amateur astronomers who work with community groups are serving youth groups (almost 90%), but adult groups were also served by more than two-thirds of amateurs, further strengthening the notion that amateur astronomers reach a broad audience.

TOPICS PRESENTED IN
EDUCATIONAL OUTREACH ACTIVITIES

Amateur astronomers offer a variety of astronomy topics in their educational outreach, though most of the activities they engage in correlate highly with their own interests. Traditional observation of objects in the night sky was the most frequently chosen topic in the 2002 survey (91% of the respondents indicated this choice). Almost four-fifths expressed high interest in speaking with the public about general astronomy, seasonal sky tours, introductions to the night sky, and current events in astronomy. Not surprisingly, the most common educational tool in outreach is the telescope, used by more than 90% of outreachers, followed by hand-outs of any kind (two-thirds) and slides (half—but they are becoming fast replaced by PowerPoint presentations).

In the 2002 survey, amateur astronomers frequently used astronomy magazines, websites, star charts, popular astronomy books, and astronomy software to prepare their educational outreach. Interestingly, the Internet already ranked almost as high as the highest ranked source of information: astronomy magazines and journals (89% versus 82% for websites, a number that in 2007 likely approaches 100%). Hence, ensuring the scientific accuracy of information on the web seems to be particularly important. Fortunately, a plethora of high-quality, reliable websites for astronomy content is currently supplying amateur and professional astronomers with content information. Yet, amateurs repeatedly mentioned astronomy content knowledge and keeping up with current events as their major challenge in conducting outreach, despite the fact that amateur astronomers on the whole have considerably more knowledge about astronomy than the audiences they serve, perhaps more than they themselves realize (see Chapter 5).

AMATEUR ASTRONOMERS' WANTS, NEEDS, AND LIMITATIONS REGARDING OUTREACH

For the most part, the amateur astronomer community seems to be in search of assistance to support their outreach efforts. Amateurs stated in the 2002 survey that a variety of support may aid them in conducting more or better outreach including ready-made presentation materials, access to equipment, training in understanding audiences, better access to experts, training in presentation skills, networking, training in content areas, and a mentoring structure that provides informal and unstructured personal support.

Interestingly, three-fifths of amateurs who engage in outreach see a lack of time on their side as a major impediment, and many requests for outreach remain unfulfilled because of it. Amateur astronomers' willingness to do outreach, hence, surpasses their capacity to do outreach. Increasing the number of active outreachers could fill a void.

AN UNTAPPED RESOURCE: AMATEUR ASTRONOMERS INTERESTED BUT NOT CURRENTLY ENGAGED IN OUTREACH

Many amateur astronomers who are currently not engaged in outreach either would like to if they had time and opportunity, or would do so if they were supported in what they perceive to be their barriers for outreach. Impediments that prevent amateur astronomers from conducting educational outreach include time constraints, perceived lack of sufficient expertise or expert knowledge (but see above and Chapter 5 of this book), lack of financial resources, lack of easily accessible materials and equipment, and group support and encouragement.

PROGRAMS TO SUPPORT AMATEUR ASTRONOMY OUTREACH

An increasing awareness in the professional science education community and among a variety of organizations that support amateur astronomers has spawned more interest in supporting and recognizing amateur astronomy outreach. The following organizations and programs are among those that provide such support:

- Astronomical Society of the Pacific (www.astrosociety.org)
- Project ASTRO (www.astrosociety.org/education/astro/project_astro.html)
- Astronomical League (www.astroleague.org)
- Night Sky Network (nightsky.jpl.nasa.gov)
- Solar System Ambassadors (www2.jpl.nasa.gov/ambassador/)
- 4M Community (www.meade4m.com)
- Sharing the Universe (a National Science Foundation funded effort by the ASP and the Institute for Learning Innovation to study and support amateur astronomy outreach)

Listed above are key organizations that support amateur astronomy outreach in a variety of ways (ASP, AL, NASA, and Meade), but also specific projects that are geared towards helping the amateur community in their outreach efforts (Project ASTRO, the Night Sky Network, Solar System Ambassadors, and Sharing the Universe). Many of these projects are informing each other, grew from one another or are otherwise closely related. For instance, one of the listed programs, the Night Sky Network, was conceived and designed based on the results of the NSF-funded research reported in this chapter. The Astronomical Society of the Pacific in partnership with

NASA and the Jet Propulsion Laboratory (JPL) inaugurated the Night Sky Network in March of 2003. The program provides member clubs with outreach activities and materials, training, and access to professional astronomers, all related to NASA mission science such as solar system exploration, the study of black holes and supernovae, and the search for planets around other stars. Subsequent independent evaluation (Storksdieck, 2005) has confirmed the positive impact of providing support to amateur astronomers in their outreach efforts. Almost half of the club outreach coordinators responding to the 2005 study felt that the Night Sky Network program allowed them to better train their own members, and one-third reported an increase in the number of members involved in outreach. Three-quarters felt the resources provided by the program were very useful and that they were able to do better quality outreach with more hands-on activities.

Based on the success of the Night Sky Network, the ASP collaborated with the Institute for Learning Innovation to start a major research effort into understanding and supporting amateur astronomy hobbyist club culture broadly and systematically. Meanwhile, the Astronomical League developed a program to recognize contributions to astronomy education and public outreach by amateur astronomers (see Chapter 8 in this book) and even commercial enterprises, such as the magazines *Astronomy* and *Sky & Telescope*, now issue annual awards for outreach.

SUMMARY AND CONCLUSION

Amateur astronomers form a large group of hobbyists many of whom are regularly engaged in educating the public about astronomy and the night sky. Previously, little was known about the characteristics of these hobbyists or the extent and nature of their outreach practices. Two major studies in the last five years, one a web-based survey of amateur astronomers conducted in 2002, the other a comprehensive summative evaluation of the Night Sky Network, have shed some light on the culture of amateur astronomers and the education and public outreach they engage in.

Amateur astronomers fall into roughly three overlapping groups: research amateurs who engage in citizen science and who support formal research efforts, astronomy enthusiasts who pursue their interest and passion in astronomy mostly by learning about it, and observation oriented amateurs who focus in their hobby on observing the night sky. The latter group is often organized in one of more than 750 clubs who count about 50,000 members in the US alone. About 20% of these club members engage in some form or another in education and public outreach, serving equally the general public and school children.

Most of the outreach is closely connected to the amateur's personal interests and hence linked to observational astronomy. Star parties for the public and in schools, school presentations, participation in large scale community events, volunteering in science museums and planetariums, and information sharing between club members form the bulk of outreach efforts that reaches potentially hundreds of thousands, if not millions of people a year (solid estimates do not exist).

In recent years, interested organizations are increasingly recognizing and supporting the efforts of amateurs in their outreach activities.

Although some amateurs are concerned about their ability to relate to their audiences, Chapter 6 will discuss research on how the audiences of outreach amateur astronomers value outreach activities by amateur astronomers. In addition, while some amateur astronomers express doubts regarding their astronomy content competence, as the next chapter shows, a large majority are quite knowledgeable about astronomy, perhaps more knowledgeable than is generally supposed by amateur astronomers engaged in education and public outreach themselves. Yet, perceived lack of knowledge, or to be more precise, a fear of not knowing the answers when asked questions about astronomy during outreach events, still forms one of the barriers to more outreach by amateur astronomers; others include a lack of time, knowledge and skills on how to connect with audiences, and teaching materials and teaching strategies. While much can be done to support amateur astronomers in their outreach, as a hobbyist group that volunteers its time and effort to share their passion with the public, amateur astronomers may very well be way ahead of their peers who engage in club-oriented serious leisure activities such a master gardeners, bird watchers, aquarists, or environmental monitoring and citizen science activists.

REFERENCES

Astronomical Society of the Pacific (ASP). (2005). Amateur Astronomy Clubs and Organizations. Online database. http://www.astrosociety. org/resources/linkclubs.html

Berendsen, M. (2005). Conceptual astronomy knowledge among amateur astronomers. In *Astronomy Education Review* 4(1). Washington, DC: Association for Universities for Research in Astronomy, Inc. http://aer. noao.edu/AERArticle.php?issue=7&section=2&article=1

Feinberg, R. 2006. Personal communication, May 30.

Fraknoi, A. (1998). *Astronomy Education in the United States.* San Francisco, CA: Astronomical Society of the Pacific. http://www.astro-society.org/education/resources/useducprint.html

National Science Board (1998), *Science & Engineering Indicators – 1998.* Arlington, VA: National Science Foundation (NSB-98-1)

National Science Board (2000), *Science & Engineering Indicators – 2000.* Arlington, VA: National Science Foundation (NSB-00-1)

National Science Board (2002), *Science & Engineering Indicators – 2002.* Arlington, VA: National Science Foundation (NSB-02-1)

Stebbins, R. A. (1992). *Amateurs, Professionals and Serious Leisure.* Montreal, QC: McGill-Queen's University Press.

Stebbins, R. A. 2006. *Serious Leisure: A Perspective for Our Time.* New Brunswick, NJ: Aldine/Transaction.

Storksdieck, M. (2005). *Planet Quest Toolkit and the Night Sky Network.* Annapolis, MD: Institute for Learning Innovation.

Storksdieck, M., Dierking, L.D., Wadman, M., & Cohen Jones, M. (2002). *Amateur astronomers as informal science ambassadors: Results of an online survey.* Technical report, Annapolis, Maryland: Institute for Learning Innovation. Available at http://www.astrosociety.org/educa-tion/resources/ResultsofSurvey_FinalReport.pdf.

5

Knowledge of Astronomy among Amateur Astronomers

Marni Berendsen and Martin Storksdieck

Now that we've discussed who amateur astronomers are and the outreach they do, are they really qualified to be informal astronomy educators?

The enthusiasm of amateur astronomers for their hobby is certainly obvious and infectious. They often provide their communities with the ONLY opportunity of putting an eye to a telescope and seeing the Moon up close for the first time, knowing it's not just a photo. This first time act of looking though a telescope often engenders an excitement that cannot be contained: "Hey, Dad! You GOTTA see this!" In addition, this experience often engenders curiosity that leads to questions: Why are there craters on the Moon? How far away is the Moon? Why does the Moon change shape during the month? How does it stay up there? Why do eclipses happen?

As other celestial sights are experienced, more questions arise. Children are especially likely to ask anything about space they are curious about: What's a black hole? Can we travel there? Where does the Sun go at night? Amateur astronomers are asked to answer these questions, and many do, providing their audiences with science information based on the audiences' immediate interest. But are their answers scientifically sound, are they "correct," or do amateur astronomers misinform the public? If we are to consider these passionate hobbyists as science ambassadors and public educators, and if we are to consider them a serious resource of informal science education, can we trust their expertise and the quality of their answers? In short, do amateur astronomers have sufficient knowledge to convey accurate information or are they, for the most part, guessing at the answers to questions from the public? Are some more knowledgeable than others? Do they know more than the public they serve? Or are they just good at operating a telescope? With all the questions thrown at them, how well prepared are they to answer accurately?

HOW WELL DO AMATEUR ASTRONOMERS UNDERSTAND ASTRONOMY?

Currently, there is no study that could answer these broad questions, though an in-depth investigation on amateur astronomer outreach culture that includes knowledge tests and structured observations will commence soon with support from the National Science Foundation. However, in a study on a representative sample of amateur astronomers doing outreach, published in *Astronomy Education Review* in 2005, one of us (M. Berendsen) found that amateur astronomers, in general, possess a high level of knowledge about astronomy. In fact, on average, amateur astronomers exhibited a level of astronomy knowledge just below that of college graduates with a degree in astronomy, physics, or astrophysics.

The Berendsen study used a standard assessment of basic astronomy knowledge, the Astronomy Diagnostic Test 2 (ADT2), developed by the Collaboration for Astronomy Education Research (CAER). More than 1,100 people responded to the online survey. Nine hundred of those classified themselves as participating in astronomy outreach. Besides basic demographics, the respondents were also asked to report on their years of astronomy club membership, their level of astronomy outreach, what informal astronomy education they have had, and their level of formal astronomy education.

The ADT2 is a 21-question multiple-choice cognitive assessment tool that covers basic astronomy concepts, such as "What color are the hottest stars?", "When the Moon appears to completely cover the Sun (an eclipse), the Moon must be at which phase?" and, "As viewed from our location, the stars of the Big Dipper can be connected with imaginary lines to form the shape of a pot with a curved handle. To where would you have to travel to first observe a considerable change in the shape formed by these stars?"

For more details on the Berendsen study and the complete online survey, see the *Astronomy Education Review* article: http://aer.noao.edu/cgi-bin/article.pl?id=137. The ADT2 diagnostic does not only cover observational astronomy, but also basic astrophysics and thus tests knowledge that tends to extend well beyond what would be needed to answer typical audience questions during a star party.

A comparison of findings from Berendsen's study ("Berendsen") with prior research that utilized the ADT2 diagnostic knowledge test (designated with a source of "ADT2") shows that college professors and those with a degree in astronomy, physics, and astrophysics scored highest, above 90% correct answers (see Table 1). Those who have been members of an astron-

Table 1. *Scores on ADT2 Assessment*

Group	Average Score	Source
College professors	97%	ADT2
Respondents with an astronomy-related college degree	92%	Berendsen
Member of an astronomy club for more than 2 years with at least one college astronomy course	85%	Berendsen
Member of an astronomy club for more than 2 years with NO college astronomy	72%	Berendsen
Respondents with at least one college astronomy course, but who have never been astronomy club members	72%	Berendsen
Undergraduates taking a 3rd astronomy course	66%	ADT2
Self-identified as amateur astronomers who are not club members, have no college astronomy, and are not doing outreach	62%	Berendsen
Interested General Public (not amateur astronomers)	41%	Berendsen
Undergraduates before their first introductory astronomy course	32%	ADT2

Note: From "Conceptual astronomy knowledge among amateur astronomers" by Margaret Berendsen, 2005, *Astronomy Education Review* 1 (4):1-18. Adapted with permission from the author.

omy club for more than two years and have taken at least one college-level astronomy-related course on average score 85% correct on the ADT2. On average, any self-identified amateur astronomer will exceed the knowledge of a member of the general public. Amateur astronomers who were not members of a club and who did not conduct educational outreach scored 62% correct on the ADT2, comparable to undergraduates who took a 3rd class in astronomy (62 and 66% respectively). Contrast this with the general public (the control group in Berendsen's online study) which scored 41% on average, or undergraduates before their first introductory course in astronomy who scored 32% on average.

The ADT2 is a rough indicator for astronomy knowledge and understanding, and there is no agreed-upon level of achievement on the ADT2 that indicates that a respondent could be considered proficient or reliably knowledgeable in astronomy. Still, the results show that most amateur astronomers, and certainly many of those who are interfacing with the public, seem to have knowledge of astronomy that resembles that of persons with considerable formal training in astronomy. The findings, thus, suggest that informal astronomy education or outreach conducted by amateur astronomers is likely to be based on scientifically sound information. Berendsen (2005) concluded further from these results that:

> The data suggest that the club experience itself, independent of formal astronomy education, has a positive effect on conceptual astronomy understanding. Except for those with an astronomy-related degree, the longer an amateur astronomer has spent in a club, the more understanding he or she gains of astronomy concepts. ...Club membership does not entirely compensate for a lack of formal education, but it does appear to give one an edge, indicating the impact of associating with like-minded enthusiasts (Discussion and Implications section, ¶ 3).

These findings are provocative since they insinuate that, by itself, taking a college-level astronomy course provides one level of understanding, but being a member of an astronomy club provides an added or additional level of astronomy knowledge and understanding. The hobbyist's passion seems to translate into informal astronomy learning for amateur astronomers, and the informal learning itself appears to be as effective as formal training. Although further research is needed to unveil the exact mechanisms by which amateur astronomers learn astronomy, the following could be contributing factors:

- Clubs schedule regular observing sessions that involve a number of members. Amateur astronomers who belong to a club are usually involved in these regular observations of the sky. Over time, they will learn to navigate the sky, predict the apparent motions of objects due to their orbits and due to Earth's rotation, and understand phenomena such as eclipses and meteor showers. Fellow members share information and help novices come up to speed more quickly.
- Generally, introductory astronomy college courses provide background in basic knowledge of astronomy and astrophysics, but lack the regular, sustained observation of the sky. Since they do not provide a personal link for the student, knowledge might be lost quickly after taking the course.
- The combination of observing phenomena while having other enthusiasts as mentors, and applying the knowledge gained through a college-level course to what is observed brings a higher level of understanding than either one of these activities alone.
- Being immersed in a culture of gaining and sharing astronomy knowledge provides a powerful and personal incentive for learning. In a supportive club environment, members act simultaneously as learners and teachers in informal teaching-learning exchanges.
- Self-study will likely explain to a large degree the high level of astronomy knowledge exhibited by amateur astronomers. The passion that these hobbyists bring to their serious leisure provides the motivational force for self-directed study.

ARE KNOWLEDGEABLE AMATEUR ASTRONOMERS WORKING WITH THE PUBLIC?

Berendsen (2005) found that long-term club members are the group of amateur astronomers most likely to be engaged in education and public outreach (see Figure 1).

Amateur astronomers in her sample who have never belonged to a club were much less likely to be participating in any outreach than those who had recently joined a club (less than two years of club membership), and recent club members were less likely to engage in outreach than established club members with two or more years of astronomy club membership. The dynamic of this relationship is not surprising: More than half of those in Berendsen's sample who never belonged to a club also never conducted outreach, a portion that declined to somewhat more than a quarter for recent members and further declined to one in nine for established club members. Conversely, less than a third of the non-members conducted oc-

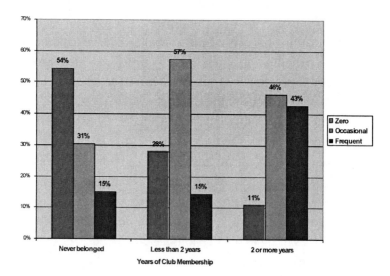

Fig. 1. Effect of Club Membership on Level of Outreach
Note: "Frequent" is defined as seven to more than 50 outreach events per year; "occasional" is defined as one to six outreach events per year. From "Conceptual astronomy knowledge among amateur astronomers" by Margaret Berendsen, 2005, *Astronomy Education Review* 1 (4):1-18. Adapted with permission from the author.

casional outreach, while more than half of recent members did. Once club members became established, the incidence of frequent outreach dramatically increased in the sample, from about 15% to 43%. Since long-term astronomy club members were also the most knowledgeable, it is more likely that the public will have contact with an amateur astronomer who will be knowledgeable in astronomy and who will therefore most likely provide scientifically correct astronomy information.

Interestingly, there was relatively little difference in formal astronomy training between frequent and occasional outreachers (see Figure 2): astronomy club members involved in *frequent* outreach were somewhat more likely than those involved in occasional outreach to have taken some college astronomy courses (55% versus 45%), and were consequently somewhat less likely to have had *no* formal training in astronomy at the college level (36% versus 45%). Formal training in astronomy does not clearly characterize the frequent outreacher as well as length of membership in astronomy clubs. As stated earlier, the study indicated that the ***combination*** of club membership and at least some college-level astronomy made a significant difference in the scores on the ADT2 assessment.

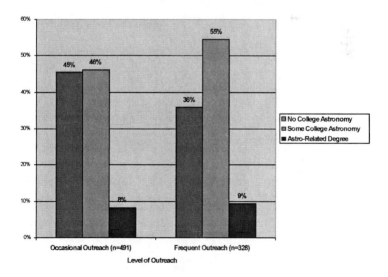

Fig. 2. Formal Astronomy Education of Club Members Involved in Outreach

WHAT TRAINING IS NEEDED AND HOW DO
AMATEUR ASTRONOMERS GET IT?

What prevents willing amateur astronomers from conducting education and public outreach and does a lack of (perceived) astronomy knowledge hold some amateur astronomers back from participating in outreach? A 2002 ASP survey on amateur astronomers and their outreach activities (Storksdieck, Dierking, Wadman & Cohen Jones, 2002) found that about half of those amateur astronomers who were not engaged in outreach stated that they simply lacked the time for it. Besides that, the survey found that 46% of those interested in outreach felt that they lacked the necessary expertise or specialized knowledge needed to conduct education and public outreach.

This finding is supported by the Berendsen study which asked respondents in an open-ended question "In what areas do you feel training would be valuable for amateur astronomers involved in education and public outreach?" Among the 430 respondents answering the question, fully 55% cited areas related to astronomy content knowledge, such as information that would be covered in an introductory college astronomy course: relative cosmic distances, explanations of observed motions or changes in the sky, an understanding of star formation, and the origins of solar system and universe.

Fear of not being able to answer the questions posed by the public, apparently, holds many amateur astronomers back from becoming involved in sharing their passion. Increased knowledge of astronomy may increase those amateur astronomers' confidence in their ability to engage with the public, and thus may increase their likelihood in becoming engaged in outreach. [However, the gap between the knowledge that some amateur astronomers might *perceive* as necessary to engage the public and the amount of knowledge truly *needed* may differ from person to person. It will also depend on an individual's willingness to say "I don't know": their willingness to acknowledge that they don't have an answer to a question].

In Chapter 7, several amateur astronomers report their journey to becoming involved in their hobby and in outreach. In many instances, they found that joining an astronomy club exposed them to a broadened understanding of astronomy, the night sky, and telescopes. It also opened their eyes to areas where their knowledge of astronomy was spotty. This inspired them to educate themselves further.

HOW DO AMATEUR ASTRONOMERS EDUCATE THEMSELVES ABOUT ASTRONOMY?

Of the 900 amateur astronomers involved in outreach in Berendsen's survey, 59% had at least some college-level astronomy training: 9% held an astronomy-related degree and 50% had taken one or more college-level astronomy courses. More than half (55%) of the respondents who self-identified as amateur astronomers stated that they had some *informal* training in astronomy, physics, or astrophysics. The degree of informal training in astronomy rises with increasing levels of club membership, outreach, and formal astronomy education (see Figure 3). For example, 69% of those who reported engaging in outreach "frequently" (more than six times a year) stated that they received informal training in astronomy, compared

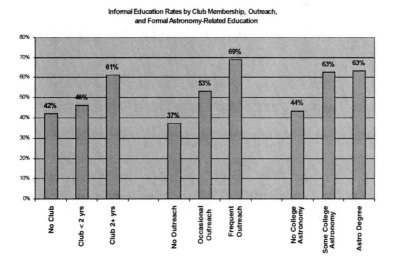

Fig. 3. Informal Astronomy-Related Education Rates

to 53% for those who engaged in outreach occasionally and 37% for those who did not do outreach. Similarly for club membership: longer membership meant higher rates for informal training. Interestingly, rates for informal astronomy education were the same (63%) whether a respondent received an astronomy-related college degree or simply took some courses. Those without college education in astronomy reported significantly lower informal training rates as well (44%). Formal and informal astronomy education seem to complement each other.

The most important sources for informal astronomy learning were identified in an open-ended question on Berendsen's survey as books and magazines (reported by 40% of the 603 respondents to the question), membership in an astronomy club (24%), and taking short classes or workshops (16%). Thirteen percent cited learning as a result of pursuing astronomy as a hobby. Another frequently mentioned category was "self-taught" (17%), which one could assume involves reading and/or research of some kind. Figure 4 summarizes the results of these overlapping categories. Taken together, the responses indicate that the majority of amateur astronomers who perceived themselves as learning astronomy in some free-choice or in-

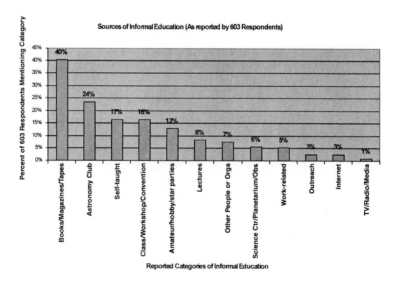

Fig. 4. Sources of Informal Astronomy Education

formal way did so in self-study, with books and magazines, less so with the Internet and TV or radio. The second important category seems to be mentoring and direct learning from others, mentioned through categories such as "astronomy club", "amateur/hobby/star parties" or "other people." Some non-formal means also played a role: workshops, classes and lectures.

How much difference, though, does pursuing informal astronomy education make? Is there a difference in the scores on the ADT2 depending on the combined level of formal and informal astronomy training?

Berendsen cross-tabulated the ADT2 scores for established club members (at least two years of membership) between formal and informal training and found that informal training had no effect on those amateur astronomers who had at least some college education—both groups scored at 85% (see Table 2). However, informal training seemed to have had a small effect on ADT2 scores for those without formal college astronomy education: Those who stated that they had some informal training scored slightly higher than who did not (75% versus 69%). Informal training, at least that which respondents identified as such, seemed to have benefited primarily those with little formal training (and thus little presumed prior knowledge), though it needs to be kept in mind that length of club membership, an indicator for exposure to informal knowledge sharing, had a strong effect on ADT2 scores!

Table 2. *Average Scores on ADT2 Assessment for Amateur Astronomers with at Least Two Years of Club Membership*

	AVERAGE SCORE	
Level of College Astronomy for Outreachers who have belonged to Astronomy Clubs for two or more years	YES Informal Astro Education	NO Informal Astro Education
Some College Astronomy	85% (n=212)	85% (n=91)
No College Astronomy	75% (n=128)	69% (n=102)

SUMMARY

Little research has been conducted on amateur astronomers' knowledge of astronomy and astrophysics. We relied heavily on a study by one of the authors (M. Berendsen) and we are quite aware that one study alone is insufficient to generalize to such a heterogeneous group. Still, the study provides us with some preliminary insights: Amateur astronomers seem fairly knowledgeable about astronomy and astrophysics, and they gain their surprisingly deep understanding and knowledge through a combination of formal training and informal or free-choice learning. Many amateur astronomers have at least some formal training in astronomy at the college level. However, college for many has been long ago, yet many seem to have kept up with astronomy through self-study, using primarily books and magazines, and through mentoring, conversations, sharing, and the almost passive absorption that occurs within a tight community of practice: the amateur astronomy club.

We posed the initial question whether amateur astronomers who engage in education and public outreach are likely to impart a factually correct body of knowledge to the public. Judging from the existing data we conclude that this is possibly or even likely so, but that more direct research is needed to investigate this question.

RECOMMENDATIONS

So what does all this mean to you, the reader? This chapter and the previous one have discussed the needs of amateur astronomers to support their outreach and factors affecting their level of knowledge about astronomy.

Based on this research, we make suggestions for amateur astronomers, for astronomy clubs, and for community members and organizations interested in working with amateur astronomers.

To Amateur Astronomers:

To increase knowledge of astronomy as well as confidence when working with the public, the following recommendations are made:

- Get involved with an astronomy club.
- Enroll in at least one college-level introductory astronomy course.
- Read an introductory astronomy textbook.
- Subscribe to and read astronomy-themed magazines.

To Astronomy Clubs:

Become the go-to source for astronomy events in your community. To actively support your club members in their outreach or their interest in becoming involved:

- Create an "Outreach Coordinator" position on your Board of Directors. This person makes the arrangements with community groups requesting astronomy events and encourages new volunteers from your club membership.
- Consider joining an organization like the Night Sky Network to acquire materials tailored for amateur astronomy outreach and receive training on presentation skills. Use these materials in regular outreach training sessions for club members.
- Provide outreach mentors: for outreach events, experienced members should invite interested newer members to accompany them.
- Hold regular training and discussion sessions for members on astronomy topics of interest to them.
- Provide recognition for your members who volunteer to work with the public.
- Contact astronomy instructors at your local junior college. Provide the class with opportunities to observe with your club. Encourage your members to take an introductory course.

To Community Members and Organizations:

If you engage amateur astronomers as volunteers, speakers, or classroom presenters, to assess the likely level of knowledge of your volunteers, ask:

- Have you ever belonged to an astronomy club?
- How long have you been a member?
- How many college-level astronomy or physics courses have you completed?

If the person has belonged to an astronomy club for at least two years, and has completed at least one college-level astronomy or physics course, you are more likely to have a knowledgeable volunteer, although there are exceptional people that do not meet these criteria. Just be aware that the person may need some additional training and support. In any case, you will have an enthusiastic volunteer and enthusiasm goes a long way toward inspiring interest in science.

REFERENCES

Berendsen, M. (2005). Conceptual astronomy knowledge among amateur astronomers *Astronomy Education Review* 1 (4):1-18. Retrieved March 1, 2007, from http://aer.noao.edu/figures/v04i01/04-01-02-01.pdf

Storksdieck, M., Dierking, L.D., Wadman, M., & Cohen Jones, M. (2002). *Amateur astronomers as informal science ambassadors: Results of an online survey.* Technical report, Annapolis, Maryland: Institute for Learning Innovation. Available at http://www.astrosociety.org/educa-tion/resources/ResultsofSurvey_FinalReport.pdf

6

The Impacts of Amateur Astronomers Engaged in Education and Public Outreach

Michael G. Gibbs and Daniel Zevin

Amateur observers have always played a major role in astronomy.
… Amateur work today is as valuable as ever, and to a considerable extent professional researchers depend upon it.

— **Patrick Moore, from the 1995 book *The Observational Amateur Astronomer***

A few [amateur astronomers] *labor at what amount to unpaid careers as research scientists, a situation that raises the question of just what constitutes an amateur.*

— **Timothy Ferris, from the 2002 book *Seeing In The Dark***

[Supernova] *2006jc occurred in galaxy UGC 4904, located 77 million light years from Earth in the constellation Lynx. The supernova explosion, a peculiar variant of a Type Ib, was first sighted by* [Japanese amateur astronomer Koichi] *Itagaki, American amateur astronomer Tim Puckett and Italian amateur astronomer Roberto Gorelli.*

— **University of California Berkeley Press Release, April 4, 2007**

The fact that amateur astronomers are important contributors to the science of astronomy is undisputed. From discoveries of new asteroids, comets, and supernovae to exceptional astrophotography and photometry, their achievements as "citizen scientists" are well recognized (Percy & Wilson, 2000). We also know that many amateur astronomers regularly volunteer their time to share their knowledge, passion and enjoyment of the sky (Ferris, 2002; Storksdieck, Dierking, Wadman, & Jones, 2002). Education

57

and public outreach (EPO) by amateur astronomers takes place in both formal and informal education settings, such as visiting a classroom or providing star parties for the public under the night sky with audiences of all ages. But unlike their contributions to science that have been documented widely in the scientific literature, their contribution and impact as informal educators to the public's understanding of, and attentiveness to astronomy are less well documented.

This chapter summarizes some of the research to date on the positive impact amateur astronomers are making as informal science educators (e.g., planetarium and science center educators—basically all individuals who teach or communicate science other than K–12 teachers, college/university instructors and professors, and professional development providers). Secondly, the authors make recommendations for further research to: a) better document amateur astronomers' EPO activities; and b) inform the development of appropriate support mechanisms to promote increased or improved amateur astronomer EPO.

SUMMARY OF RESEARCH TO DATE

Amateur astronomers, on average, are exceptionally knowledgeable about astronomy (Berendsen, 2005) and they regularly take part in EPO activities (Storksdieck, et al., 2002). But what is the real impact they are having as informal educators? We know of no study to date that has formally tested student knowledge (or that of any other audience) before and then again after interacting with one or more amateur astronomers. However, evidence from both the formal classroom and informal worlds of education suggests that amateurs are making valuable contributions.

Formal education

Project ASTRO is one of the best-known classroom programs that officially recognizes and enlists amateur astronomers as volunteer educators. Developed at the Astronomical Society of the Pacific (ASP) in 1993, under the leadership of then executive director Andrew Fraknoi, Project ASTRO is a national program that links professional and amateur astronomers, scientists, graduate students and other similar volunteers (collectively referred to from this point forward as volunteer astronomers) with teachers and their students (primarily in the fourth through ninth grades) through regional sites throughout the United States. The program's main goal is to develop long-term partnerships between participating volunteer astronomers and teachers to bring accurate and engaging astronomy learning into the classroom (Fraknoi, 2000). Participating professional or amateur astronomers typically visit their "adopted" classroom at least four times each academic year with the goal of inspiring students' interest in science through hands-on and inquiry-based astronomy activities. An integral component of Project ASTRO is an intensive two-day astronomer-teacher training workshop that takes place before the astronomer visits the classroom and interacts with the students. During these workshops, held annually at each participating regional site, emphasis is placed on blending the teacher's knowledge of instructional methods and classroom management with the astronomer's knowledge of and passion for astronomy and science. Each year, over 200 volunteer astronomer-teacher partnerships are newly formed through Project ASTRO and almost 2,000 such partnerships have been created to date. It should be noted that amateur astronomers comprise over half (57%) of the volunteer astronomers participating in the program. The remainder are categorized as either professional astronomers (24%), college students (12%), or professional astronomy educators (7%), a category that includes anyone who is engaged in astronomy EPO via a museum, science center, planetarium, or other informal science education institution (ASP, 2007).

Project ASTRO was piloted in Los Angeles and in the San Francisco Bay Area in 1993, with support from the National Science Foundation (NSF). Based on the success of the pilot project, the ASP received funding in 1995 from NSF and NASA's Office of Space Science to expand Project ASTRO to several other regions and it is now offered through 11 regional sites around the country. Both the pilot project in California and the national expansion of Project ASTRO were independently evaluated by the Institute for Learning Innovation (ILI).

The summative evaluation of the pilot project focused on teacher satisfaction and perceived changes in the classroom as a result of participating in Project ASTRO. The results of the investigation suggest that participating volunteer astronomers had a positive impact in the classroom. For example, 91% of 54 teachers responding to ILI's post-project survey reported teaching more astronomy as a result of participating in Project ASTRO and 48% even reported teaching more science in general. When both teachers and volunteer astronomers were asked whether or not their students had "actively learned astronomy" as a result of their involvement in Project ASTRO, 68 (74%) of 92 respondents felt that their students had, while 20 (22%) answered "somewhat" and 4 (4%) gave no answer. In addition, 75 (81%) of the responding participants felt students were "more excited about astronomy" as a result of the program, while 15 (16%) said that their students were "somewhat" more excited about the topic. When participating teachers were asked to rate their overall Project ASTRO experience on a scale from one to five (one meaning "least useful" and five "most useful"), 52 (97%) of the 54 responding teachers rated their experience as either a four or a five. When broken down by the type of volunteer astronomer partnered with the teacher (in this case, either "professional," "amateur" or "astronomy educator"), 61% of those teachers partnered with an amateur astronomer rated their experience as a five, compared to only 38% of those partnered with a professional astronomer and 20% of those partnered with an astronomy educator (Dierking, 1994).

Components of the summative evaluation of the national expansion of Project ASTRO focused mostly on the ASP's ability to successfully replicate the program at other sites across the country, while a supplemental "Student Impact Study" took a look at the program's impacts on participating students. This occurred through written questionnaires and interviews. The data from this investigation suggested that students may have been influenced in the following ways as perceived by teachers, parents, and students themselves (Anderson, 1998):

1) Project activities had positively affected students' attitudes about science and astronomy;
2) Project activities had increased students' knowledge of astronomy concepts and ideas;

3) Project activities had increased students' confidence in doing science; and

4) Project activities had contributed to students' understanding of the activities in which professional and amateur astronomers engage in.

In a more recent study the ASP, with assistance from the Project ASTRO National Network, conducted a national survey of in-service teachers participating in Project ASTRO. In-service teachers were asked to comment on the effectiveness of Project ASTRO in improving their students' attitudes toward science and correcting common misconceptions in astronomy or science. Gibbs (2006) reports that of 177 in-service teachers participating in the on-line survey, 44% indicated the program had a large positive impact on their students. Thirty-eight percent stated they felt the program made a slight improvement and 8% indicated there was no change. Regarding the ability of the program to correct common student misconceptions in astronomy or science in general, 91% indicated "yes" (51% "yes, very much," and 40%, "yes, somewhat"). Eight percent responded "not sure." One percent responded "not really." Additionally, 60% of in-service teachers reported that they taught more science as a result of Project ASTRO.

A study by Gibbs and Berendsen (2007) indicated that in-service teachers participating in Project ASTRO perceived amateur astronomers to be at least as effective as professional astronomers in inspiring their students' interest in science, and that Project ASTRO amateur volunteers may have been especially effective at the elementary school level. For example, when the elementary schoolteachers in the survey were asked about change in their students' attitudes toward science in general as a result of their participation in Project ASTRO, 77% of those who partnered with an amateur perceived a "large positive change," in contrast to 50% of those who partnered with a professional astronomer.

The research regarding Project ASTRO provides some preliminary support for the idea that amateur astronomers are having a positive influence on both teachers and students in the formal world of classroom education. But what about amateur astronomers' impacts in the informal world of education, otherwise known as "free-choice learning," where attendance and participation are—to a large degree—optional for the audience?

Informal education

Astronomy education is not limited to just the classroom. It also occurs in planetariums, museums, nature centers, and at star parties, where amateur astronomers (often organized through their clubs) provide telescope viewing opportunities for the general public (Ferris, 2002; Storksdieck, et. al. 2002). It is estimated that at least a few hundred thousand children, families and adults participate in such events each year (ASP, 2007). These informal astronomy learning events, well-known in the amateur astronomy community, are becoming increasingly better coordinated and publicized—even internationally. For example, on May 19, 2007, the first-ever "International Sidewalk Astronomy Night" resulted in hundreds of amateurs' making their telescopes available for stargazing to passersby on city streets the world over (Gay, 2007).

But there is more occurring at star parties than people peering through telescopes. Inaugurated in 2004 by the ASP and NASA, the Night Sky Network (NSN) is one of the most recognized national programs that support the public outreach endeavors of amateur astronomy club members. Through the NSN, the ASP in partnership with NASA has developed and tested a number of activity-rich toolkits of which hundreds of copies have been distributed to astronomy clubs across the country to assist their members in conducting informal outreach learning events. Each toolkit focuses on a particular astronomy topic (e.g., black holes or the solar system) and includes demonstrations and hands-on activities for the public related to NASA research programs. Membership in the NSN is limited to astronomy clubs in the United States and one toolkit per topic is provided to each club. The program also encourages participating amateurs to share stories and to learn from one another through a community-oriented website (http://www.astrosociety.org/education/nsn/nsn.html).

Although there is to date no comprehensive research that we are aware of on how public events conducted by amateur astronomers are contributing to their audiences' enjoyment and appreciation of astronomy, there is self-reported feedback and anecdotal evidence from the NSN to indicate that amateur astronomers' informal education activities do provide their audiences with a unique and exciting experience while encouraging them to become more engaged in and attentive to astronomy in general. For instance, NSN members provided the following quotes while logging their EPO activities online:

The most exciting part of this star party was when three teenage boys rode up on bicycles and looked through my scope. They were really interested and stayed a long time, asking tons of questions. — *Member, Astronomical Society of Northern New England*

The most surprising part was how many parents stayed for the whole thing. We were at the library and some parents who had dropped their kids off and were enjoying the library decided to come to the presentations. We gave pre- and post-tests and saw a substantial increase in basic solar system knowledge (from 19%–49%). There is a real need for more education and public outreach to help supplement what the kids are learning in school. — *Member, Southeast Ohio Astronomical Society*

This event provided many opportunities for informal teaching to a non-captive audience. We did several activities for kids who had missed our previous sessions. Many sat in for the review because they had had so much fun the first go-around. After only three previous observation sessions I was very pleased to see that they remembered the main constellations and planets from The Spring Sky. Awesome! My favorite part? They keep coming back for more on their own...I think they are hooked! Just shows the power of direct experience! — *Member, Norman North Astronomy Club*

[At a viewing of the Mercury transit] One woman struck up a conversation with me and asked several questions about the solar system, galaxy, etc. She seemed well-informed but was glad to have someone to verify facts and to share information with. When the clouds finally cleared and she saw the solar surface and finally tiny Mercury she was thrilled. Next I noticed she was telling passersby to come take a look at Mercury. Some club members had more than one scope set up and she was more than happy to help newcomers to take a look and to tell them how to find Mercury. She was tickled to be able to share with others something that had so impressed her. — *Member, Whatcom Association of Celestial Observers*

In addition, amateur astronomers reported via web-based surveys that the toolkits were effective with their audience and the support provided by the NSN enabled many to expand or improve their outreach activities (Storksdieck, 2005). The assumption is, therefore, that participants leave a star party or other types of amateur outreach events inspired and motivated. Additionally, participants may continue to learn about astronomical topics on their own after such an event, thereby fulfilling one of the main objectives of free choice learning: a lifelong, personally motivated pursuit of knowledge and understanding (Falk & Dierking, 2002).

Beginning in 2007, a new National Science Foundation funded study being conducted by the ASP and ILI will, among other things, test this assumption by examining the educational value, as perceived by the audience, of the EPO activities of amateur astronomy clubs. By looking at audience engagement with a variety of outreach activities offered by amateur astronomers and by measuring audience self-reported satisfaction with these events, potential changes in their attentiveness to astronomy as a result of such events, and their perceived and (to a lesser degree) actual knowledge gained through the events, this new project (titled *Sharing the Universe*) will for the first time document in the astronomy education literature some of the genuine impacts being made by amateur astronomers in the informal education world (Storksdieck pers. com., 2007).

Another informal astronomy education program that enlists amateur astronomers is Family ASTRO. Family ASTRO was initially developed in 2000 by Andrew Fraknoi at the ASP. With National Science Foundation funding, Family ASTRO has expanded to six regional sites and is reaching diverse audiences through a series of hands-on activities, discussions, and games facilitated by event leaders that include, among others, amateur astronomers (ASP, 2007).

In 2004, ILI completed its final summative evaluation report of the Family ASTRO program. Some of the major conclusions from the ILI report were as follows (Ellenbogen, Foutz, Haley-Goldman, & Adelman, 2004):

- The National Family ASTRO Team created a successful model for developing high quality science activities for families. Both event leaders, and families were overwhelmingly positive about their experiences.
- Of the 128 event leaders surveyed for the summative evaluation, 22 (17%) considered themselves to be an amateur astronomer.
- Participants expressed a consistently high interest in participating in an event again. Follow up interviews with parents conducted three to six months after an event showed that they remained interested in attending an event again and that their rating remained at the same high level that it was during the event. In addition, more than a third (36%) of the 531 surveys collected at events indicated that families had attended at least one other event in the past.
- Family ASTRO was used collaboratively in the home. In the months following their attendance at a family event, almost three quarters (74%) of 49 interviewed parents reported that someone in the family had done a Family ASTRO take-home activity since the event. Of those, 81% reported that parents and children did the activity by working together.

The Family ASTRO program thus provides additional evidence that amateur astronomers, as presenters in facilitated programs, may be quite capable of connecting with their audiences in positive ways and meeting many of the goals of informal learning, such as inspiring audience members to be more attentive to astronomy in general, to be more engaged in astronomy activities, and to seek additional information on their own after participating in astronomy-related events.

CONCLUSIONS

During the past several years a number of reports were issued on problems with science education and science literacy in the United States, such as

Rising Above the Gathering Storm from the National Academy of Sciences (NAS, 2007). But it did not take a government study for amateur astronomers to know there is a problem, for they have known for years there are common misconceptions among the public with such basic science concepts as astronomical distances and sizes, the objects that make up our solar system, and even the general use of telescopes and how they operate (Berendsen, 2005). Better still, amateur astronomers are out there, competent in their subject matter (Berendsen, 2005), reaching hundreds of thousands of people each year doing their best to dispel misconceptions in astronomy (ASP, 2007). The evidence to date indicates they are having a positive impact in both the formal classroom setting and in their communities. Through programs such as Project ASTRO, amateur astronomers are bolstering in-service teachers' confidence in teaching astronomy and students' attitudes about science in general. Through their own outreach and engagement activities, they are increasing the public's interest in and general attentiveness to astronomy, and presumably, are motivating people to learn on their own.

Further research is needed to gain a better understanding of the contributions amateur astronomers are making in education, and to gauge in more detail what support they need to increase and improve their ability to conduct EPO activities. Future research into amateur astronomers' EPO activities could focus on the following:

1. Document in more detail, more specifically, and with more rigor amateur astronomers' contributions to education;
2. Conduct more independent evaluations of all new projects or significant project expansions of current projects that enlist amateur astronomers for education purposes (see Bailey & Slater, 2005; Stroud, Groome, Connolly, & Sheppard, 2007) for recommended best practices for astronomy education evaluation);
3. Investigate the most effective techniques and methods used in amateur astronomy (e.g., should astronomy clubs focus events in areas where they are likely to interact with a higher number of people, such as a mall parking lot or a city sidewalk, or is it best to be located in

dark/remote locations where viewing is exceptionally better?);
4. Research into impacts that show quantifiable instead of only perceived knowledge gain by those involved (e.g., looking at student
test results before and then again after interacting with a volunteer
astronomer partnered with a teacher).

Overall, evidence suggests that amateur astronomers have the potential
to enhance and expand the public's knowledge of astronomy and space
science. Amateur astronomers should therefore continue to be encouraged
and supported in their EPO activities. Those who are professionally employed in astronomy and space science and in education in general can
play a large role in this by supporting and otherwise giving more attention
to partnerships between professional educators and amateur astronomers.
In addition, through further support and expansion of amateur astronomy
clubs' EPO endeavors, amateur astronomers' influence as informal educators will only continue to grow.

REFERENCES

Anderson, D. (1998). Project ASTRO: Student Impact Study. Results of
an evaluation conducted for the ASP under NSF award ESI-9552551.
Annapolis, MD. Unpublished report.

Astronomical Society of the Pacific (ASP). (2007). Education program
databases, unpublished data.

Bailey, J.M. & T.F. Slater. (2005). Finding the forest amid the trees: tools
for evaluating astronomy education and public outreach projects. *The
Astronomy Education Review*, 3(2):47-60. Retrieved April 2, 2007 from
http://aer.noao.edu/cgi-bin/article.pl?id=120

Berendsen, M. (2005). Conceptual Astronomy Knowledge Among
Amateur Astronomers. *Astronomy Education Review*, 1 (4):1-18.
Retrieved April 2, 2007, from
http://aer.noao.edu/figures/v04i01/04-01-02-01.pdf

Dierking, L.D. (1994). The Astronomical Society of the Pacific's Project
ASTRO: Summative Evaluation. Results of an evaluation con-

ducted for the ASP under NSF award ESI-9253156. Annapolis, MD Unpublished report.

Ellenbogen, K., S. Foutz, K. Haley-Goldman, & L. Adelman. (2004). Final Summative Report: Family ASTRO. Results of an evaluation conducted for the ASP under NSF award ESI-9901892. Annapolis, MD. Unpublished report.

Falk, J.H. & L.D. Dierking. (2002). *Lessons Without Limit: How free-choice learning is transforming education.* Walnut Creek, CA: AltaMira Press.

Ferris, T. (2002). *Seeing in the Dark.* New York: Simon and Schuster, 379 pp.

Fraknoi, A. (2000). An Overview of Project ASTRO. *Mercury,* 29(1), 18.

Gay, P.L. (2007). *SkyandTelescope.com Homepage News,* May 14, 2007. Retrieved June 29, 2007 from http://www.skyandtelescope.com/news/home/International_Sidewalk_Astronomy_Night.hl?showAll=y&c=y

Gibbs, M. (2006). Taking Steps to Make a Difference. *Mercury,* 35(5), 26.

Gibbs, M. & Berendsen, M. (2007). Effectiveness of Amateur Astronomers as Informal Science Educators. *Astronomy Education Review,* 2(5):114-126. Retrieved April 2, 2007 from http://aer.noao.edu/cgi-bin/article.pl?id=228

Moore, P. (1995). *The Observational Amateur Astronomer.* New York: Springer-Verlag, Berlin Heidelberg. 280 pp.

National Academy of Sciences (2007). *Rising Above the Gathering Storm: Energizing and Employing America for a Brighter Economic Future.* Washington, D.C.: The National Academies Press.

Percy, J.R. & J. B. Wilson (Eds.). 2000. Amateur-Professional Partnerships in Astronomy, ASP Conference Proceedings, Vol. 220.

Sanders, R. (2007). Massive star burps, then explodes. *UC Berkeley News,* © UC Regents. http://www.berkeley.edu/news/media/releases/2007/04/04_supernova.shl.

Storksdieck, M. (2005). Planet Quest Toolkit and the Night Sky Network. Annapolis, MD: Institute for Learning Innovation. Unpublished report.
———. (2007). Personal communication, June 25.

Storksdieck, M., Dierking, L.D., Wadman, M. & M. C. Jones. (2002). Amateur Astronomers as Outreach Ambassadors: Results of an Online Survey prepared for the ASP under NSF Planning Grant ESI-0002694. Annapolis, MD: Institute for Learning Innovation. http://www.astrosociety.org/education/resources/AAISASurveyResults.pdf.

Stroud, N., M. Groome, R. Connolly, & K. Sheppard. (2007). Toward a methodology for informal astronomy education research. *The Astronomy Education Review*, 5(2):146-158. Retrieved March 1, 2007, from http://aer.noao.edu/cgi-bin/article.pl?id=229

7

Women Hold Up Half the Sky

Judy Koke

WHAT'S THE PROBLEM?

Women who came of age during the time of Mao Tse-tung were taught that "women hold up half the sky." This expression has been used frequently since that time: as a title for planetarium shows featuring female astronomers, mental health programs targeted to women, and as a motto for many National Women's Day events. Yet women represent only a small percentage of astronomy club membership and are underrepresented in science, technology, engineering, and mathematics (STEM) majors and careers both in the United States and in most industrialized countries around the world (GOA, 2004). The science community became quite interested in this situation in the late 1980s, when it was recognized that although girls and boys perform equally well in these subjects in elementary and middle school, by high school girls demonstrated a marked decrease in confidence and interest in math and science. Many organizations, such as the National Science Foundation, made this issue a priority and, as a result, a vast number of "girls in math" or "girls in science" programs have been funded and implemented across the country. Many of these programs do excellent work and, in fact, girls' math and science SAT scores, as well as undergraduate participation in STEM, has increased over the past ten years. However, women continue to be underrepresented as STEM graduate students or professionals.

One might be tempted to question if this discrepancy is indeed a problem. As a researcher in this area I have been asked more than once if it really matters that men seem to prefer STEM careers while women want to be teachers. Isn't it more important that we do what makes us happy, even if that might mean different things for men and women? The answer is that the situation is more complex than that. The reasons that many educators, scientists, and policy makers consider the current gender gap a problem derive from concerns for both the fields and the individuals. First, and perhaps most importantly, science is shaped by the questions it asks. As the questions we ask are rooted in our frames of reference (hence, strongly shaped by our background and culture) men and women tend to ask different questions and, in fact, ask them differently. As scientists, mathematicians, and engineers build their, and hence our, understanding of the world, the ability to see both questions and answers from many perspectives will only help make scientific explanations more complete, relevant, and useful. In other words, the inclusion of women and minorities in STEM work and research improves the quality of STEM fields themselves. Secondly, many very talented and intelligent women and minorities are choosing to contribute to other areas, which might be considered a significant loss for STEM fields. Finally, as a matter of equity, we want to ensure that any woman, in fact any person, who does wish to consider STEM fields is truly welcomed and able to succeed if they so desire (Blickenstaff, 2005).

Currently women represent approximately 15% of amateur astronomy club membership, (Storksdieck et al., 2002) and this author wanted to explore why that might be. How do interested women come to connect with astronomy? If a women is interested in astronomy, does the environment of most amateur astronomy clubs welcome and support her? What do women think would draw more women to the pursuit of amateur astronomy? To explore these questions, a web survey was developed and sent to several astronomy club contacts, with the request that they share it with any female amateur astronomers they knew. Based on years of experience, and the reliance on word-of-mouth and professional connections for distribution, this author expected approximately 75 responses in the two weeks the survey was posted. Much to my surprise some ten times that number of women responded, from countries as distant as Australia,

Egypt, and Japan. From these 781 survey responses we are able to glimpse the world of amateur astronomy through the eyes of interested, committed women everywhere.

This chapter first reviews what research says about gender differences in STEM learning, which can inform, if desired, how astronomy club activities are developed and implemented in the future. Secondly, we'll examine what women had to say about astronomy and their experiences in astronomy clubs. Finally, we consider some possible recommendations for how amateur astronomy clubs might engage more women.

WHAT RESEARCH SAYS ABOUT WOMEN AND STEM LEARNING

Interestingly, boys and girls perform equally well on standardized tests of science and math in elementary school, which suggests that biology may *not* be at the root of the gender gap in STEM. Research has explored the biological differences between men and women for over a century, identifying such differences as relative brain size (men's are slightly bigger), connection between hemispheres (women's are better connected), and areas of the brain utilized to perform specific tasks (Legato and Tucker, 2005). In the early 1990s, Dr. J. Hyde conducted a number of interesting studies which are summarized in a paper detailing the physiologic (or structurally based) differences in learning (Hyde, 1996). Although she was able to document a small but significant difference in spatial perception, her findings did not account for the limited participation of women in, for example, engineering fields. In fact, it is clear from the research literature that whatever biological differences exist between men and women, there is very little difference in their scientific or mathematical abilities, and certainly not enough of a discrepancy to explain the under-representation of women in STEM careers or amateur astronomy clubs.

One major difference that has been identified is women's *attitudes* towards STEM, which become significantly less positive than men's attitudes through the teenage years. One indicator often used to illustrate this attitudinal difference is "confidence," as girls begin to underestimate their abilities in middle and high school. However, these findings are complicated by the fact that "bragging" is behavior that is strongly associated with boys and considered highly unfeminine and undesirable among young women. Other studies suggest that girls have significantly fewer hands-on experiences in science as they are often engaged with lower frequency with tools, machinery, equipment, and computers. Many researchers suggest that the shortage of women pursuing degrees in STEM is due to this earlier lack of preparation, experience, and mentoring (GOA, 2004).

In addition, there is certainly decreased enrollment of girls in high school physics, chemistry, and higher math courses. Girls tend to see "higher math" as unnecessary for everyday life. Girls see basic math as the most important of all subjects in middle school; described as necessary for basic life skills such as getting a mortgage, balancing your check book, and so on. Girls tend to see higher math as necessary only if you want to be an engineer or scientist, which are already perceived by girls in middle school as not "family-friendly" and hence not female occupations: Young women perceive college-level science courses and science-related careers as isolated, inflexible, competitive, and incompatible with motherhood (Holland and Eisenhart, 1990). Women continue to major in and earn degrees in STEM to a lesser extent than do men, even though women now comprise the majority of all college students (GOA, 2004).

Research has also documented what helps girls be *successful* in math and science. Dr. Sue Rosser's *Principles of Female Friendly Science* (Rosser, 1990) suggest that girls need to:

- Understand how a topic is relevant to society
- Have additional or "catch-up" experience in hands-on tinkering with equipment and tools
- Learn in a collaborative (noncompetitive) environment
- Have strong mentors; adult modeling helps to overcome reluctance to get dirty, take risks (i.e., do something "wrong"), or persist at a task rather than give up or get help

Young women like STEM better when it is seen as "helping" or contributing to society. They need to see the "so what" of what they are learning and discovering. Science and higher math are usually presented as the pursuit of abstract ideas, which only sometimes generates information relevant to our everyday lives. Young women want to understand how communities use science everyday, to "know what they know" in order to make informed decisions, and to successfully negotiate life on the planet. They need to "see" the opportunities for positive social impact that a science career might offer (Blenky, 1997). They need to practice STEM skills in collaborative, inclusive environments.

It is important to note that a focus on creating programs for "girls in science" is not as exclusionary of boys as one might think. In a large study conducted in Germany on physics-related interests of 8,000 boys and girls (ages 11–16) attending different types of schools, researchers determined that "although girls and boys in some domains have a somewhat different interest structure, there is a considerable overlap in interest" (Haussler and Hoffman, 2002). Moreover, what is interesting for girls is almost always interesting for boys, but not necessarily vice versa.

What does this mean for astronomy? Women are good observers, an essential skill in astronomy. However, they are often intimidated by their lack of experience with the technology utilized in astronomical explorations. Research suggests that they need to be in an environment where it is safe to ask the most elementary questions, and that they prefer to participate in very collaborative, social projects. Perhaps most importantly, the pursuit of astronomy as a profession or a hobby needs to be perceived as compatible with caring for a family.

WHAT DO WOMEN SAY ABOUT HOW THEY BECAME INTERESTED IN ASTRONOMY?

When asked how they originally developed their interest in astronomy, 17% of respondents explained that their interest stemmed from being introduced to astronomy by a parent or grandparent as a youth. Many reflected eloquently about the aura of "specialness" they felt—the mystery of being out at night, camping or at home in the yard, and looking up at the stars with a parent. In fact, another 14% of respondents felt it was the sky itself that had called them to astronomy.

My Dad would sit out every night and watch the stars. I would sit with him as he would point out the stars and constellations. I was hooked!

I used to star watch from my back porch in Appalachia, where it sometimes seemed you could fall into the sky.

In addition to the 14% of respondents who felt their interest in astronomy originated in the night sky, an equal number of respondents could not pinpoint the source of their interest, but rather conveyed that they had always felt connected to astronomy.

I have loved astronomy since I was a very small child.

Born with it.

Another 11% of respondents became interested in astronomy through a class they had taken at school, at the elementary, high school, or even college level. Eight percent said they had become interested because of a spouse's or friend's enthusiasm for astronomy, while 7% became interested through a book they had read or through a camp program, planetarium, scouting group or museum (7%). Twelve individuals specifically named Carl Sagan as the source of their interest, either through TV specials, books or movies. Additional sources of interest included astronomical events such as the Apollo program or the appearance of Halley's comet, and the need to learn astronomy for their job.

When I was eight I read my brother's Boy Scout book on astronomy and it fascinated me. I tried to learn everything I could by reading all the books I could get my hands on. I got my first star finder by taping coins to a piece of cardboard and sending it an address [sic] *in a* Family Circle *magazine.*

Watching the Apollo missions as a child.

College astronomy got me hooked, but I've always been amazed by the goings on in the universe.

I became interested because my husband had a telescope that he made himself. He showed me how to use it and I thought it was fun.

I saw Carl Sagan on the Phil Donahue *show. Carl was talking about his book* Broca's Brain. *I read everything Carl wrote and as a result fell in love with astronomy and science.*

The answers to these questions were often long and rich with personal details that chronicled the paths these women had taken in their study of astronomy. Many suggest a dogged commitment to study the subject even when formal schooling was not accessible. The responses suggest that these women, both astronomy club members and not, harbor a deep and abiding passion for astronomy, which, put to targeted use, could well serve as a contagious catalyst for other young women.

WOMEN IN ASTRONOMY CLUBS

Fifty-eight percent of the respondents were current members of at least one astronomy club at the time of the survey (May 2007). The remaining 42% were not, but may have been previously. Although not asked directly, a small number of respondents volunteered that they belonged to multiple clubs.

When asked to describe their main reason or purpose for joining an amateur astronomy club, women's responses indicate that they join to learn. Fully 35% of respondents cited as their reason for membership a gain in knowledge about either astronomy as a subject (25%) or in their ability to utilize the technology now used to study astronomy (10%).

It has good speakers and can improve my knowledge of astronomy.

I've owned a telescope for nearly 10 years and I never really learned how to use it properly. I was hoping to meet more experienced astronomers and learn from them.

Another 24% joined an astronomy club to participate in a community with shared values and interests.

I wanted to be able to socialize with people who have the same passion that I do. It's very disheartening when you see something awesome in the sky, and you're all excited about it. Then you go tell someone, expecting them to be just as excited, and their answer is something like, "Aaa, yea. So?"

I just wanted some camaraderie and learning. I'm not interested in "girl" talk.

I wanted to be a part of a group of people that loved astronomy just as much as me.

The other reasons women joined astronomy clubs were to access equipment or places (9%), to find safety for dark sky viewing in remote places (8%), to share pre-existing expertise or knowledge with others (7%), to join an interested friend or spouse (6%), to support their careers (2%), or to meet intelligent people (or even men!) after a move (2%).

I was the liaison between a science museum and the astronomical society. I also used it to further my knowledge.

I am a full time parent of two young children and need intellectual stimulation of this sort regularly!!!

Safe observing sites and observing companions

Almost two-thirds (65%) of women feel their club meets or exceeds their intended goals and needs. Only 6% were truly disappointed in their club, while 28% offered a qualified endorsement of their club's ability to meet their needs:

Relatively well, however, it would be nice to have more people in the group who were my age.

There are some terrific folks with whom I have made a good and meaningful connection. There are others in the club that are insufferably sexist and controlling.

More than one third (34.2%) of the clubs that these respondents belonged to (clubs were of all sizes) have fewer than 10% female membership;

another 29.4% of the clubs include 10–20% women. Only 14% of the clubs consisted of a third to half women. Two clubs were described as 60–69% female. This means that women participate in clubs where they are largely outnumbered by male co-members, and that astronomy club membership does not reflect community demographics: astronomy clubs attract men far more than women. This sample of women astronomy club members reflects a wide distribuition of club membership duration, as 25% of respondents had been members for each of the participation-length categories: 1) less than 2 years; 2) more than 2 but less than 5; 3) more than 5 but less than 10; and 4) more than 10 years.

When asked if there is a difference in how men and women participate in their clubs, 37% of women said there was indeed a difference, while a significant number of additional women suggested there were too few women in the club to observe a pattern. This difference was described as being related to men's stronger interest in the technical tools (often referred to as gadgets) and to the fact that many women are less comfortable at dark sky star parties, due to their often remote and dark locations, and the lack of amenities.

> *The balance of work and family obligations is big for women—having a family makes it more challenging to get out to the meetings. The more women there are the more support we became to one another. The idea of hanging out in the dark observing with a bunch of men sounded a little unusual/unsafe/odd. In the beginning it seemed safer and more comfortable to have at least one other female with me. (I know both myself and my husband felt more comfortable knowing that I was not the only woman). I have found there are a decent number of women who attend the meeting—but rarely the group observing.*

Women were also asked how they would describe the attitude of their club towards women. Eighty percent felt their clubs were welcoming or very welcoming, and a number of respondents responded negatively to the very nature of this question. However, 15% of women suggested that their clubs were somewhat awkward for women, while only 5% felt they were actually hostile towards them. Respondents did not feel that there were many barriers to women taking a leadership role in their organization; in fact, many respondents had served as club presidents, treasurers, or other officers.

It is very welcoming, but the fact that there are so few women involved is a bit awkward to younger women.

Welcoming, but since there are sooooo many men, it could seem like a "guys only" club, just because that's who's there.

When asked how astronomy clubs might better serve women, individuals were highly prolific, responding with long, detailed, and thoughtful suggestions. Many participants suggested that clubs should not focus on simply encouraging females to join, but rather to increase the diversity of the clubs by thinking about age, race and ethnicity as well. It was suggested that both current club demographics, and the fact that astronomy can be an expensive hobby if one considers the technology involved, were barriers that need to be overcome. One respondent quoted the US census at length and then concluded "*If clubs are not reaching out to minorities, they are on the slow path to decline.*" Another respondent went on to explain that "*diversity is the key to both the success of astronomy clubs, and of astronomy in general.*"

Reinforcing this concept that broader diversity was crucial, the most frequent suggestion (22%) in response to the query of how clubs could better serve women was to develop special interest groups, or hold special interest events, with a focus on young women, introductory level skills, or youth.

I have wanted to start a women in astronomy organization that would help women astronomers band together and not feel so isolated…a teaching organization, but also a bonding organization.

I think that many women would like to get into the hobby, but may be intimidated by a lack of technical knowledge as well as being insecure in their ability to handle a telescope. I think we need to help these women to gain their skills and confidence by having beginner workshops.

I think it needs to start young...there [is] also need to be strong role mod-
els. They would have to be women that are knowledgeable and passionate
about astronomy. They can't be seen as the "planners" of the group. Also,
I know that girls like to be social and help others, thus I think it would be
good if the club was more than just a hobby. For instance, our club often
supports local scouts obtain [sic] their astronomy badges. We often go to
schools and hold special events for the public. I really enjoy these types of
activities and I think other girls would also.

Numerous recommendations to improve recruitment strategies were
the second most frequent response category (13%), suggesting that clubs
should actively focus on recruiting female members. A number of individuals suggested that many women are not aware that astronomy clubs
exist, which was further supported by responses from the nonmembers
described later in this chapter.

Advertise them more. I think that there are many women out there who
would like to know about astronomy clubs, but do not know they exist.
Clubs need to go TO the women.

Through marketing and community outreach by showing women, girls
and families they can be physical scientists...and that they can have inter-
ests and hobbies alongside men.

Another repeated suggestion was to make astronomy clubs more "family
friendly" to encourage women to come with their children, or to organize
a way to provide child care. Women continue to hold primary responsibility for childcare in our society, and including families in some club activities would support larger or more frequent participation from the young
mothers. The last question of the survey requested any additional information that respondents would like to include. A surprising number of wom-
en articulated that they wished men would understand that for women in
particular, frequency of participation does not correlate with dedication,
as many times women are hampered from participating because of family
care issues. The topic of safety and comfort at dark sky observing events

mentioned above was also reinforced in the answers to this question, with a particular emphasis on bathroom facilities:

Maybe by having special activities for their kids. Our club does a lot of educational things for kids, but there aren't many kids at the observing sessions.

When we had our "Mars Watch" the advertisement said to bring your children, so I think that helped women to know they were welcomed.

In our club it seems women are more inclined to attend meetings, programs and social gatherings like observatory open houses, so these events should be encouraged. I also think [it] helps to set up more "formal" type observing sessions. Our club has a couple of nights a month set aside at our dark sky sites. The challenge is that some women in the club tend to stay away from "informal" sessions unless they are sure that someone else they know will be there. They understandably don't want to show up alone and find that no one else decided to show up that night.

Another major consideration in choosing dark-sky sites is that the guys do not worry about restrooms, and that is a very important consideration for me. I have always been the lone female advocating restrooms.

The topic of recruiting female speakers was also a lively one. Respondents suggested that not only could those female speakers serve as role models for women and young people, but also that female speakers often speak about astronomy in a way that is appealing to other women.

…we've had very good speakers who happened to be women, which has helped demonstrate that women are as welcome in the club as men.

There needs to be a general shift towards more female speakers, actively encouraging female members and a culture of disapproval of sexism—or any other kind of discrimination.

Give women the opportunity to report on their particular astronomical interests. I have been fortunate that I have been given these opportunities, and feel respected for my specific astronomy interests.

They need women speaking to women about the technical aspects of tele-scopes and observing. Men tend to use technical terms assuming everyone else knows what they are talking about.

Finally, many female members feel slighted when treated in stereotypi-cally ways, such as being personally asked to bring food to events, and pre-fer these requests be made as an open one to all club members.

WOMEN WHO DO NOT BELONG TO CLUBS

Of the 42% of respondents who do not belong to an astronomy club, just under one third indicated that they would like to but can't find one close to home. This is either because they have searched but were unable to locate one in their area (18%), or they are simply unaware of a specific club or clubs in general (11%). The most frequently cited reason for not belong-ing to a club, however, was lack of time (27%). Many women explained that they would join a club if they could, but the responsibilities of family, school, and work were prohibitive. Other reasons included a sense of in-timidation (3%), based on either lack of knowledge or the predominance of male members, a lack of funds or equipment (5%), or a negative experi-ence with clubs in the past (6%).

Non club-members have adapted by finding many alternative resources to support their passion for astronomy. They rely heavily on the Internet (35%), hardcopy publications such as books or periodicals (15%), or a more knowledgeable friend or family member (15%). Still others find pro-grams at local schools, planetariums or parks, or star parties offered to the general public.

Nonmember respondents' general impressions of amateur astronomy clubs is overwhelmingly positive, with the most frequent negative impres-sions being the predominance of males, experts, or geeks (in that order), and an emphasis on technology and "gadgets." Only 15 individuals report-ed that they prefer to participate in astronomy as a solitary endeavor. All of these findings indicate that should amateur astronomy clubs choose to make an effort to engage more women in their membership, those efforts should be richly rewarded.

A ROLE FOR ASTRONOMY CLUBS IN THE FUTURE

A small number of respondents to this survey expressed strong irritation with its focus on women and difference. Those very questions were shaped by my original research question: Why do women make up only 15% of astronomy club membership? If I had chosen a different question, or if someone else had approached the same question, this survey and its findings would have looked markedly different. Much as the questions I asked shaped what little bit more we now know about astronomy clubs, the questions scientists choose to ask—or not ask—in astronomy shape the entire field. Astronomy as a field, and STEM more generally, will all be stronger fields if they are more inclusive and diverse.

But how can we support that change? What supports diversity in college level STEM enrollment and graduation? We know that more frequent hands-on experience and practice with STEM tools and skills, combined with strong mentors and role models and an understanding of how STEM contributes to society, will make a gradual, but definite improvement. So what can astronomy clubs do?

Generally we can surmise that amateur astronomy clubs are welcoming organizations where a passion for astronomy is shared and fostered. However, many women feel hampered in their desire to participate due to the issues described above. Astronomy clubs should make a conscious effort to:

• More frequently engage families in observing activities;
• Actively recruit and welcome female membership;
• Ensure a balance of female speakers and role models;
• Decrease the barriers to participating in dark sky observing events;
• Develop female-friendly approaches to teaching technology skills; and
• Underscore contributions to social issues

Astronomy clubs will gradually see their club membership change, until a critical mass of women are members, and the issue will no longer require conscious effort because change will continue organically. These changes will, in all likelihood, increase diversity in age, economics and ethnicity as well, helping to ensure a stronger future for both the club and the field.

ACKNOWLEDGEMENTS

The author wishes to acknowledge the assistance of Dr. Laura Danly, of the Griffith Observatory, and the Astronomical Society of the Pacific, in the dissemination of this survey.

REFERENCES

Blenky, M., et al. (1997). *Women's Ways of Knowing.* New York: Basic Books.

Blickenstaff, J. (2005, October). Women and science careers: leaky pipeline or gender filter? *Gender and Education,* 17:4, pp 369-386.

GOA Report: GOA-04-639: *Gender Issues.* (July 2004). National Center for Education Statistics, Post Secondary Institutions in the United States: Fall 2000 and degrees and other awards conferred 1999-2000, (Washington, DC, National Center for Education Statistics, 2001.

Haussler, P., and Hoffmann, L. (2002). An Exploration of Student Understanding and Motivation in Nanoscience. Available at: http://hi-ce.org/presentations/documents/Hutchinson_etal_NARST_07.pdf

Holland, D. and Eisenhart, M. (1990). *Educated in Romance: Women, Achievement and College Culture.* Chicago: The University of Chicago Press.

Hyde, J.S. (1996). Meta-analysis and the psychology of gender differences, in B. Laslett et al., *Gender and Scientific Authority.* Chicago: University of Chicago Press.

Koke, J. (2005, Spring). "A Lifetime Investment: How Adolescent Girls View Science Careers" *ASTC Dimensions.*

Marianne J. Legato and Laura Tucker, (2005). *Why Men Never Remember and Women Never Forget,* New York: Holtzbrink Publishers.

Rosser, S. (1990). *Female Friendly Science: Applying Women's Studies*

Methods and Theories to Attract Students, Pergamon Press.

Storksdieck, M., Dierking, L.D., Wadman, M., & Cohen Jones, M. (2002). Amateur astronomers as informal science ambassadors: Results of an online survey. Technical report, Institute for Learning Innovation. Available at http://www.astrosociety.org/education/resources/ResultsofSurvey_FinalReport.pdf

8

In Their Own Words: Amateur Astronomers Tell Their Stories

Marni Berendsen

Research into amateur astronomy outreach is a relatively new undertaking. Prior chapters have reported the available research on the practices, knowledge, and effectiveness of amateur astronomers engaged in outreach.

This chapter brings this research to life as we feature the voices of four amateur astronomers who talk about their personal experiences of sharing astronomy and the night sky with the public:

Joan Chamberlin
of the Astronomical Society of Northern New England in Maine

Wayne "Skip" Bird
of the Westminster Astronomical Society in Maryland

Dave Rodrigues,
the "AstroWizard" at the Chabot Space and Science Center in California

Rosemarie Spedaliere
of the Astronomical Society of the Toms River Area in New Jersey

Each has his or her own style, knowledge, and areas of interest. These imaginative and passionate people share what it's like to spark in their visitors a sense of awe and wonder about the universe. As volunteers, why did they choose this path, what was it that sparked their own interest? What impact do they see they are making? What are the challenges they can face when holding a public event? How do they feel supported in doing their outreach?

Through their own words, let's meet these amateur astronomers whose stories represent the thoughts and feelings of astronomy enthusiasts who share their passion with the public.

SPARKING A LIFE-LONG PASSION

The path to becoming an amateur astronomer and going on to sharing the sky with their communities can start with a curiosity about a special event such as the arrival of a bright comet, a supportive parent pointing out constellations, or a milestone in the history of space exploration such as the first man on the Moon. But often it comes from an encounter with an enthusiastic amateur astronomer.

Dave:

I have been interested in astronomy and astronautics ever since I was four in 1957. As a child I was fascinated by the Space Race, then in full flower. What was so wonderful about it was that as a child, I couldn't differentiate between reality and fiction, which made it especially exciting. So, I couldn't tell if we had already landed men on the Moon, how big a space station was, or if we had already gotten to Mars, or what exactly was going on. The Space Program in those days felt like the night before you go to Disneyland as a child. In some ways the anticipation is even more wonderful than the realization because you could project your imagination and fantasies upon it.

A pivotal point in my life occurred when I visited Chabot Space and Science Center one Saturday evening in the '60s, when I was in seventh grade. Kingsley Wightman, the former director of Chabot Observatory, towering above me (or so it seemed), looked down at me, pointed at me, and said "You, young man, should join the Eastbay Astronomical Society [EAS]." I got the application and joined.

I cajoled my parents into taking me to Chabot once a month on Saturday to EAS meetings. I was amazed and astounded! I was able to talk to real engineers, scientists and astronomers! There, I fell under the influence of Terry Galloway. Terry was fresh from getting his Ph.D. in Chemical Engineering at Caltech. He would talk with passion about Feynman's

Physics books. I had never met anybody like that. And he was so friendly and encouraging to me and many other young people at Chabot.

Skip:

I have always had an interest in the sky, day or night, but it wasn't until I went to Hawaii to Mauna Kea in November of 1985, that I got the bug. The father of a friend was involved with the research there. He was going to be in Hawaii at the same time I was so he invited me up to see the observatories. I got to see Halley's Comet and more stars than I had seen even living in the mountains of Colorado.

A few years later, in 1993, I had just moved to a new house in Westminster, Maryland which was near a park, where the Westminster Astronomy Club was having a star party. My wife, Phyllis, and her mom were taking a walk to look at the park and came back to tell me there were a bunch of people with telescopes down there. I remember one guy who was there named Bruce Wrinkle. He had a 6-inch F-12 Astrophysics refractor. After looking through every telescope from his refractor to a 20-inch Obsession, I thought, "Now this is the way to do this hobby!" So I joined the club.

Our club became involved with outreach when Comet Hyakutake came for a visit in 1996. The members had all decided to meet up at Bear Branch Nature Center where our club meetings are held. In a small town, word gets around. Someone told someone who told someone else and the next thing we know we have 200 to 300 people assembled at the Center to see the comet. It was exciting to have so many people interested in what was going on above their heads. Once you have the first kid look through your scope and say "WOW COOL," you're hooked. Enthusiasm is contagious. About ten of those people showed up at the next meeting wanting to look through our scopes and become club members.

Joan:

My romance with astronomy began with the Space Race, but I never had a real astronomy course or even looked through a telescope until much later in life. When I was a Girl Scout leader, I taught myself the bright stars and the constellations using books by H.A. Rey. After a year of looking at stars, it was a thrill to see a star come up in the east that I hadn't seen for many months and to call it by name. It was like greeting a friend.

My obsession with astronomy had begun, and I couldn't wait to share it with my Girl Scout troop. My first real outreach event was a star party our troop organized in a field at one of the girl's homes out in the countryside. It seemed to me that the girls did as much running around in the dark as actually looking at the constellations I was pointing out, but just a year ago I met one of those girls at a supermarket parking lot in the city. A couple of Greater Portland Astronomical Society members and I were out there in the freezing weather with our telescopes. I heard someone call my name. It was my former Scout, Carolyn. I hadn't seen her in ten years. She said, "I knew it had to be you. Remember when you showed us the constellations at Anna's house and on our camping trips? You didn't have a telescope then." It was such a pleasure to show Carolyn the moon and Saturn through my telescope. It was the first time she had looked through one. I was so touched to think that those nights I spent showing the stars to my troop had actually meant so much to her. What better reward could there be for doing outreach?

Rosemarie:

As a young girl growing up in the '50s and '60s I loved science and especially astronomy. I was not encouraged except for my mom who took me on regular trips to the Hayden Planetarium in New York City. She also bought me my first book on the constellations when I was six. I was also fortunate to attend a high school that offered a course in astronomy, which I took in my senior year.

In the year 1996, I purchased my first telescope. It was a Meade 60mm refractor. The only thing I could find with it was the Moon. I was disappointed. A few years later, I attended Astronomy Day festivities at the Robert J. Novins Planetarium and the club members of the Astronomical Society of the Toms River Area (ASTRA) were there with their telescopes set up on the side of the hill. They were more than welcoming and shared their telescopes to show me how to find some of the Messier objects and Saturn. That night I joined the club. Of course I went home and had to see Saturn in my own telescope and found it all by myself! I was so grateful to the members who helped me with celestial navigational techniques and who gave me telescope tips. It was an experience I will always remember.

I soon became very involved in the club and when I found out that we could apply for Night Sky Network membership I did so immediately. I wanted to share the night sky with the public and, like ASTRA had inspired me, to help them to dust off those long-forgotten telescopes in their attics and enable them to marvel at the wonders of the Universe. Having a background in public relations, I was compelled to excite the public and stir their interest in astronomy, the hobby I loved so much.

IMPACT OF OUTREACH

The rewards of working with the public range from inspiring by entertainment—that science can be exciting and surprising—to a sense that you are making a difference in the way people think and the paths they take in life. Amateur astronomers see the value especially through the eyes of children, letting them discover that science is fascinating, fun, and for everyone.

Joan:

Astronomy outreach is important for students and for the public. When students can actually look through a telescope at an object and connect this to what they are learning about astronomy in the classroom, their experience becomes more real. It is much more exciting. Many teachers have not had lots of training in astronomy and amateurs are a great resource. When amateurs pair up with students to do actual astronomy projects, the learning is even greater. Astronomy outreach can inspire students to go into careers in math, science, or engineering. Another astronomy group that I have become active in is Southern Maine Astronomers. In addition to other activities, the group works with high school teachers and students on photometry projects [which involves measuring the brightness of stars]. I am working on becoming more knowledgeable in photometry to improve my astronomy knowledge and to be able to help students with these projects. Photometry is an exciting way to experience scientific research first hand.

The public can benefit from outreach events too. It's surprising how little the public knows about what's up there in the sky or how it works. I know because I used to be part of that public. I do believe most of the public prefers to be knowledgeable and they seem to enjoy learning. I've had many people thank me for my presentations and say how much they appreciated learning it. I once did a private viewing session for an 80-year-old woman. She had always wanted to look through a telescope at the Moon or a planet, but she had never had the opportunity. It was wonderful to see the expression on her face at her first view of the Moon. Everyone deserves to have seen the Moon or Saturn through a telescope.

One of the outreach events that I am most proud of is the Space Day event I organized for the six schools in my school district in 2004. It was the largest event I have ever organized and it took courage to take it on. I was fortunate to have the support of the Northeast Regional Coordinator of Space Day, Sharon Eggleston, a passionate and creative person who believed I could do it. This was a very special event for my school district in rural Maine. I involved the businesses and community members of all five towns. Many businesses decorated their shops with a space theme. One of the restaurants had their workers dress as Martians. Community members read space books in the elementary classrooms. We had a four-mile scale model Solar System that passed by the largest elementary school, with a brochure that community members and students could take on their tour explaining the planets. The entire school district met for a presentation by a retired astronaut and proclamations from the governor and state senators.

All classes had presentations that week with the NASA moon rock and meteorite samples. We had a main speaker, Sam Kounavas, who talked about the Mars Rover mission. Students participated in a poster contest that advertised the evening star party for the communities. Several amateur astronomers provided solar viewing. Every class had at least one classroom presentation. I found 35 presenters for that day, which included two Solar System Ambassadors, a TV meteorologist, college professors, a female pilot, and amateur astronomers from the groups that I belong to. That evening we had 12 amateur astronomers at the school for a public star party.

The support I received from everyone from school lunch workers and custodians to the superintendent of schools was terrific. I have been doing Space Day for our district ever since, although it's a bit scaled down from that first one.

Last year I did a Space Day presentation for a fourth-grade class that I had done a different presentation for the previous year. When I asked the class how many students would like to become scientists, six or seven students raised their hands. I really believe that it's important to have special events like this to have younger children begin thinking about careers in science.

I continue to do outreach because I am passionate about astronomy; I love sharing it; I believe in inspiring students to enter math, science, and engineering; and I believe if the public is more knowledgeable about astronomy and space, they will be willing to fund science missions. I want science missions to be funded because I want to know more about what's up there!

Rosemarie:

I personally have had so many wonderful experiences doing outreach. I will never forget the time a girl of about 12 years of age came to my scope and said, "Can I look?" "Sure," I said. She asked whose telescope it was and I told her it was mine. She said "But you are a girl, can girls have telescopes?" I said, "Of course" and pointed out the women in our group and their telescopes. I also told her not to ever let anyone tell her girls can't like science and astronomy. At that point she turned to her parents and shouted, "See, I CAN be an Astronomer!" Needless to say it was very rewarding.

We have intellectual discussions and crazy ones! Recently a young man asked, "If we are in the Milky Way, how do we get pictures of it?" Who can ever get tired of hearing someone shout "WOW" when seeing Saturn for the first time? My personal thrill is to have them watch for the International Space Station and see it. I stand there waving my arms saying "Hi Guys!" as the space station passes over us.

Then there was the time we did a star party for kids fighting cancer. The night with those kids was unforgettable. Quite a few asked us, "Can we see Heaven?" I had lost my 12-year-old daughter to cancer and it just made me feel good to be there. I was glad to be of some help to them so they could forget their illness just for the night while looking at our Universe.

In today's society where the emphasis is put on reality TV shows, rowdy talk shows, sports, celebrity lifestyles, and chat rooms we need to educate the public about "outer space" not "cyber space". Space is our future and our children's future. By doing public outreach we not only show the public some fantastic sights in the sky, we inform them about space exploration and the milestones achieved.

Skip:

One very special event our club held was the program we did for the National Federation of the Blind with a group of their campers. These kids were not only blind, but teenagers far from home. Many for the first time. It was absolutely amazing to see this group go from a very tired and frustrated looking bunch to a group hovering around a table soaking up everything we had to say.

It took every one of us to pull this off. We proved the concept that we could acquire an image at the eyepiece, process it for printing onto Swellform paper, which is then heated in a graphics machine to create a raised, tactile image, all in less than 4 minutes. It shows that we could be describing an image at the scopes and before we finished speaking, the kids could be examining the images by touch. To watch how excited these kids were to be able to "see" something out in space for the first time was thrilling. We also entertained with several different presentations, adapting to the special circumstances of working with blind students. The campers were all disappointed when the counselors finally said it was time for bed. It was a very rewarding experience to be involved with this program. You get that warm glow when you know you did the right thing and that someone's life was changed because you cared.

Dave:

In the 1960s, the Apollo program was just getting started. Our club's mentor, Terry, had the bright idea of writing NASA and asking if Chabot could be any help with the program. After all, we had these really bright high school and college kids and really big telescopes. Could we do anything useful to help?

Amazingly NASA wrote back and said YES! It's hard to believe now, but back in those days NASA could launch a spacecraft to the Moon or Mars and … MISS! Part of the reason is that calculating orbits is a process of successive approximation, requiring LOTS of computer power that we just didn't have back then. It's not much of an exaggeration to say the Casio Data Bank wristwatch that is on my arm as I write this is as powerful as was a room-sized NASA computer in the early '60s. So one of the NASA contractors asked us, along with a number of other observatories around the world, to help track the *Apollo* spacecraft and provide positional navigational data. This would act as a backup and confirmation to the Doppler radar data provided by the Deep Space Network.

In the first few missions, NASA was WAY off in their predictions as to where we would see the spacecraft. This was a relatively calm and academic task until the explosion aboard *Apollo 13*, when all at once our data became a life and death matter. I remember, as if it were yesterday, seeing the *Apollo 13* Command, Service and Lunar Module as a "slowly" moving star in the eyepiece of "Rachel", Chabot's 20-inch Brashear refractor.

So, what's the relevance of this story to Astronomy Outreach, you may ask?

One of our young club members at that time, John Bally, went on to become a Professor of Astronomy at Colorado University. Another student member, Robert Schalck, went on to help make the optics for the *Mariner*, *Viking*, and *Voyager* Spacecraft.

A 12-year-old sixth-grader visited Chabot and asked our former director, Kingsley Wightman, if he thought we'd ever get to Pluto. Kingsley thought for a moment and answered. "I think Pluto is a little too far but I think someday we'll get to Jupiter and Saturn." That young man grew up to be Dr. Torrance Johnson of J.P.L., the *Voyager* Spacecraft Imaging Scientist, who was in charge of taking the *first images of Jupiter and Saturn!* Dr. Johnson told me that story himself.

And finally, around that time, we had a 13-year-old boy in our audience. He was a seventh grader at a local junior high. The name of this gawky, misbehaving teenager? Tom Hanks. Yes, the same Tom Hanks who went on to star in the movie *Apollo 13* and made the HBO series *From the Earth to the Moon*. Tom knew more about the Apollo mission than the director of the movie, Ron Howard, did. Tom knew about the *Apollo 1* fire, the details of the *Apollo 13* accident, and even how some of the spacecraft systems worked. He had learned it from us as a child in our audience.

There is a VERY relevant moral to this story: WHEN YOU GIVE AN ASTRONOMY OUTREACH PROGRAM, YOU NEVER KNOW WHO IS IN YOUR AUDIENCE!

As an amateur astronomer giving a Star Party or doing an astronomy outreach event, you could be changing the shape of the future in ways you can't possibly imagine.

CHALLENGES TO OUTREACH

Amateur astronomers face a number of challenges in their outreach, among them are recruiting enough volunteers to accommodate the number of visitors, gaining confidence and finding ways to effectively talk to the public (especially to children), and a common dilemma: cloudy skies and cold or wet conditions. But amateur astronomers devise creative ways to handle them.

Rosemarie:

As in everything dealing with the public there will always be obstacles and challenges. It is always difficult to make the "weather cancellation call" when the group is so excited about coming. We sometimes offer a rain/cloud date to appease them yet it's hard to make everyone happy. Then there are times we don't cancel, but instead hold a program indoors using the Night Sky Network kits and some PowerPoint presentations that we put together. We also show them how mythology relates to the different constellations. We will set up a telescope and use this trick: Across the room in a can hanging from the ceiling we have a slide of Saturn. The indoor telescope is pointed at it and if you didn't know better you would think we were looking at the real Saturn outdoors.

Another of the hardest things for some amateur astronomers to overcome when doing public outreach is the fact that we do not have the perfect dark-sky conditions. Many public places have flood lights and motion detectors. The "faint fuzzies" cannot be seen through the telescope. That's OK as the general public is less interested in them anyway. We use more prominent items to observe, such as the planets. We try to schedule the star parties when there is a crescent or quarter moon as it is always a crowd pleaser especially when small children are present. Yet there are some dates scheduled that have the full moon. We don't cancel; we make the best of it.

The public tends to not dress properly for the cool weather. It is important to prepare them for that since we live in a cool climate. This is why we try to plan to have someone inside a building doing demonstrations so our visitors can warm up a bit.

Then there is always the young person who "hangs" on the eyepiece or knocks the telescope out of alignment. To avoid this, as each person approaches the telescope I give them a quick run down of stargazing and telescope etiquette. I show them where the telescope controls are and how to focus their view. I provide a step stool for shorter people. I turn the eyepiece to the side to accommodate even the youngest child. It is important to realize that we are not in a serious, personal observing session and to make it fun and exciting.

Oh and before I forget, we now always make sure there are no underground sprinklers on. One night we were in the middle of a star party at the Planetarium and the college neglected to turn off the automatic sprinkler system. To protect our equipment everyone ran quickly and stood on a sprinkler head while I ran into the boiler room to figure out the sprinkler timer and to shut it off. I am always prepared for even a pop-up shower. I keep a large garbage bag handy to throw over my scope in a suddenly wet situation.

Then there are always those lovely puffy clouds that show up uninvited just as the telescopes are set up. One night in particular we were at an elementary school. The night was pretty clear and we took the children into the cafeteria where we did some activities with them. By the time we got outside the clouds rolled in and we could only peek at Saturn once in awhile. So you make the best of a bad situation. This particular school was up north from where we live. Instead of looking at the sky, we pointed our telescopes at the New York skyline and the Empire State Building. They were so thrilled and learned that telescopes could be used for terrestrial observing as well.

Skip:

The biggest challenge when you start doing public outreach is the public is HUNGRY, hungry for more. Once the word got out [that our club gave public events], everyone wanted astronomy programs, from a local Cub Scout den to the Air and Space Museum of Washington, DC. We had to put someone in charge of coordinating the events and, unfortunately, tell some people "No," at least until we got more people who liked to work with the public. As always enthusiasm is contagious, if you get excited about it so will someone else. When one of our club members first started in outreach, it was hard to get him to say two words to a group of more than one person. He was very smart, knew all the facts and figures out to the year 2113 for any astronomical event but when it came time to try and relay that knowledge to anyone else it was like he was a brick wall.

So I would talk to the group and ask him for the numbers. Eventually he felt comfortable talking on his own. Now he does 99% of the planetarium shows each month by himself.

Dave:

It may be hard for some adult astronomers to remember how alien and complex our discipline can be to children. We are adult authority figures who tower over the kids and who may inadvertently use arcane and confusing terminology. But what child doesn't love Wizards? In the early 1990s I had to go to a costume party. I decided to go as a Wizard as they were the ancestors of modern Astronomers. Besides, I always wanted to be a Wizard! I then realized that dressing up as a Wizard for my Star Parties would make it more fun and less intimidating to children.

In 2000 I started working at Morrison Planetarium as a Planetarium lecturer on Sundays. It was great fun! The staff at Morrison let me wear my Wizard's costume during the children's shows.

One day, Kirsten Van Stone, the AstroWizard's biggest supporter on the Morrison staff, came to me and asked if I would be willing to do astronomy-themed birthday parties for the Academy. I decided to give it a try. To my amazement, birthday parties turned out to be a great venue for teaching children Astrophysics! Between my magic tricks and real science demos that *seemed* like magic, I was an enormous hit. Soon I was teaching children's classes at the Academy called "Introduction to AstroWizardry" and "Advanced AstroWizardry." These were modeled after how I thought an Astronomy class would be taught at Hogwarts, Harry Potter's Alma Mater.

The rising popularity of Harry Potter was another Godsend for me. I already fit the bill and there are lots of humorous Astronomical references in the J.K. Rowling books. Our school systems might take an important lesson from Hogwarts School: Astronomy is a required course for young wizards from their first year on. Wizards also love to name their children after stars and constellations. Characters such as Sirius Black, Draco Malfoy, Bellatrix Le Strange, and Merope (Lord Voldemort's mother as well as one of the stars in the Pleiades) are central to the stories. Children who have, in many cases, read and reread the books many times are thrilled to discover that they can see stars and constellations named after their favorite characters.

Joan:

Since my early days of outreach, I have grown and changed because of my contacts and involvement with many astronomical groups.

I joined the Astronomical Society of Northern New England (ASNNE), where I learned how to use my 10-inch Dobsonian telescope and how to find deep sky objects in the sky. ASNNE had just opened a new observatory, and they were quite active in outreach. At first I wasn't much help, but I kept going to all their events with my telescope. With the help of some wonderful members, I was soon finding objects on my own and became caught up in the fun of showing the public what's up there. This bolstered my confidence, but I still felt I needed to know more to make my outreach meaningful.

The Southworth Planetarium at University of Southern Maine began offering short astronomy courses. They were inexpensive, non-threatening, and challenging. They even included math. Taking these courses was a big turning point in how I would approach outreach. Armed with more knowledge about how things work up there and how we discovered it, I was able to tell star party visitors more about the objects they were viewing. It was a giant boost to my confidence, and the more I learned, the more I wanted to know. I took some of the courses twice just to really absorb them! I also began volunteering for outreach events at the planetarium and occasionally volunteered as a projectionist.

OUTSIDE SUPPORT FOR OUTREACH

Support for astronomy outreach comes from within the astronomy club as well as from outside organizations, such as local schools, science centers, and even from NASA whose websites are overflowing with resources. Local community organizations may provide financial support as well.

Regional, state, and national parks provide venues for astronomy outreach, at times along with darker skies. A few nationwide programs are designed to provide direct support and resources for amateur astronomy outreach: The NASA Night Sky Network (http://nightsky.jpl.nasa.gov), the Solar System Ambassadors (http://www2.jpl.nasa.gov/ambassador/), and the Astronomical Society of the Pacific's Project ASTRO (http://www.astroso-ciety.org/education/astro/project_astro.html).

Joan:

Much of the outreach I do is through the astronomy clubs I belong to and the Southworth Planetarium at University of Southern Maine. All of these groups have other members who help with the outreach too. We have a great time together, and I enjoy their company. We learn from each other and share an appreciation for the wonders of the universe. Other members are my greatest support.

For the Space Day event, I received support from the Maine Space Grant Consortium and a mini-grant from the school system. Parent Teacher Clubs offered more funding.

The Night Sky Network is also a tremendous support. The teleconferences with scientists, the toolkits, and the discussion board provide me with resources and answers to questions. I am fortunate that my husband, although he is not very interested in astronomy, is supportive in my astronomy outreach addiction.

Skip:

For a few years our club was hit or miss on public programs until one of our members, Brian Eney, mentioned something about a toolkit he had gotten from some group called "The Night Sky Network." We sat down and played with the programs and they were fun and interesting. This gave us something to do and talk about when it was cloudy or snowing which around here is more likely than clear.

Other club members started coming out and helping, when they'd see how much fun and how easy it was and they started doing some of the programs themselves. Each member will help in their own way.

Along the line of materials and support for our program, there are lots of companies from places like Lockheed Martin, Space Telescope Science Institute, and NASA, to our local science center.

Rosemarie:

I was very fortunate to have a very knowledgeable group of friends who I regularly observed with. I approached them and asked for their help with public outreach and they backed me 100 percent then and still do so today. We began to receive requests for "Star Parties." I officially became the club's "Star Party Coordinator."

My friends were very helpful and shared their expensive equipment with the general public. They were patient and understanding and willing to answer even the craziest questions. They were always available to lend a helping hand to any visitor who brought their unused telescope from home.

ENGAGING AMATEUR ASTRONOMERS FOR AN ORGANIZATION, SCHOOL, OR COMMUNITY

Now we hear some tips from our astronomy enthusiasts regarding what a community organization might expect when they request a visit by a group of amateur astronomers. The amateur astronomers also need to know a bit about the needs of the requesting organization.

Primarily amateur astronomers offer the public the opportunity to look through telescopes and experience the universe first hand, often referred to as a "star party." But they offer much more. From hands-on activities that explain the phases of the Moon or how to make a comet to full presentations on black holes or discovering planets around other stars.

What information needs to be exchanged between the astronomy group and the requesting organization for an effective, entertaining, and engaging event? Essentially, the amateur astronomers need to know how many people to expect and the age ranges, any special requirements the group has, where the event is to be located, what time of day (or night), and how much time they have. Providing enough notice ahead of the event allows the astronomy club to gather enough members so the audience gets plenty of personal attention.

Skip:

Knowing how many people to expect, and their ages, allows the club to make sure they have enough people and/or scopes for the group. There's nothing worse than fifty 10- to 13-year-old boys waiting in line to look through one scope.

I deal a lot with adjusting my presentation to the audience because I teach an astronomy class for a home school church group. You just tailor the program to only include certain subjects. We also did the program for the National Federation for the Blind where we had to adjust our *show*-and-tell to *touch*-and-tell.

Talking about location, we tried to do a star party in a parking lot of a local library, under street lights, next to a strip mall, on a busy corner. We could see maybe four or five stars (which was enough to align a scope so we could find a few things). But this was the normal night sky to some of these people and they still wanted to look. In spite of that, it got enough of a response that the library wanted to do it every three or four months.

Rosemarie:

When an organization asks us to book a date I always ask why they want us there and what they need from us. Some groups such as the Boy and Girl Scouts have badges and patches to earn. I ask about how many people and the age range. We've shared with people as young as 2 and as old as 85. I ask how long they will let us stay, what type of grounds we will be on, and if we can park our cars by our telescopes. Some dates get booked far in advance so that they can be advertised yet others are short notice. We try to accommodate them all.

The location of the event at times provides us with unexpected or unique experiences. We've seen turtles being born in the wild before a star party and have done events everywhere from beachfronts where everyone is wearing shorts and flip flops, to a mansion's poolside deck with people in tuxedos and evening gowns.

Joan:

When a community group or school requests a presentation from our astronomy club, we need to know the size of the group and whether the group is made up of a certain age group. The size of the group will determine how many telescopes are needed if it is for a star party. If it's a presentation, I may use a different type of presentation depending on the number of participants and their ages. For example, if the participants are young children, I would try to have something more active where they can either make something or have a discussion rather than a lecture. The length of the presentation would also be determined by the age of the participants.

Knowing specific needs or interests is also important because the presentation can be chosen to accommodate those interests. If students are studying about the solar system, it might be more appropriate to choose a presentation about the planets or asteroids than to choose one about supernovae.

Lead time is important because all of us have other commitments and will need time to prepare for a presentation. If a star party is planned, it's important to have a rain date. In our state, star parties often get canceled for poor skies. If the group is only available on a specific date, sometimes an indoor presentation can be planned to take the place of the star party.

Whatever event the organization chooses, they can expect amateur astronomers who are passionate about astronomy and eager to share it with them.

A FINAL NOTE

A large percentage of amateur astronomers eagerly and regularly share the universe with their communities. Not only do they provide many people with their first look through a telescope, they also partner with teachers in the classroom, volunteer at science centers, and offer workshops for youth organizations, such as Scouts and after-school programs.

The impact amateur astronomers make is not simple to measure. We can count how many events are held and how many people are reached, but the experience of looking beyond Earth and gaining a wider understanding of our universe can have a life-long impact. As one of the contributors to this chapter, Dave Rodrigues, said, "There may well be (and probably are) future scientists, engineers, politicians, writers, artists, certainly voters and, yes, perhaps even an actor listening to you and looking through your telescope. That cute little girl in the red t-shirt may some day walk on the Red Planet. That 6-year-old may be a future Nobel Laureate in Physics. The feelings I had as a child explain, in part, my approach to astronomy outreach today, especially to children. I want them to feel the same sense of mystery and adventure and, yes, Magic, about astronomy and the universe that I felt when I was their age."

Connect up with your local astronomy club. Ask them about the programs they offer. Become a member. Share the magic.

9

Putting It All Together…

Mike Reynolds

Observational astronomy amongst amateur astronomers has always been encouraged in so many different ways over the years. The accomplishments of amateurs, for example from comet discoveries to solid variable star observations to occultation timings, have been touted as examples of how amateur astronomers can submit useful observations and data. Often through these observations, or through opportunities to simply set up telescopes for general viewing, amateur astronomers have the opportunity to share the heavens with the general public.

There is an interesting and amazing statistic from the Association of Science-Technology Centers, ASTC, that more people in the United States go to a science museum or science center each year than attend professional sporting events. This is probably a surprise to many of us who see the domination of sports at all levels on television, radio, and print media. More importantly: this statistic tells us not only that the public continues to be interested in science, but more so tells us a little bit about how the public would like to engage with science—in a captivating, hands- and minds-on way that is easily understood, is playful, and entertaining. Various chapters in this book have demonstrated that amateur astronomers can provide the public with engaging science experiences in a knowledgeable fashion. The book also highlighted the many challenges that still remain when hobbyist volunteers share their passion with the public.

This book gives us an introduction into our emerging understanding and recognition of amateur astronomy outreach; the importance of astronomy outreach; and the accomplishments of both individuals and clubs/societies. This is an important step, for the amateur community can play and is playing an important role alongside both formal and informal astronomy education.

What does it take to "turn on" a child or an adult to a topic such as astronomy? For David Levy it was a meteor, Jim Kaler a star. For me it was the launch of Alan Shepard in 1961; I was simply hooked and found myself living and breathing space. And then the universe was literally before all of us.

Today amateurs are on the front line as not only marketers of the universe, but as educators. As was noted previously, ponder all the times you were asked how far away the Moon is or what a black hole is. Amateur astronomers are a competent group to answer these questions. Marni Berendsen's data is indeed provocative and in many ways reassuring, as is Michael Gibbs and Dan Zevin's look at this all-important topic. Why do amateur astronomers know so much? They learn from their peers in their local clubs, study by themselves from magazines, books and increasingly more the internet, but they are also motivated to take introductory college astronomy classes. Many of the students in the college-level astronomy classes I teach are indeed amateur astronomers.

But what about the amateur astronomer in the classroom—on the teaching side of the desks. Tim Slater's overview is an important discussion for those who want to enter that arena. I often hear amateurs tell me that they feel poorly-equipped to teach in a more formal setting, even though they "know" their astronomy and drip with enthusiasm. Certainly the classroom should be much more than a straight lecture, built on a constructivist viewpoint. And with some training, many can make excellent astronomy educators; I have seen a number that have developed into excellent instructors. And some even end up in the field of formal education as a profession!

Astronomy outreach has the opportunity to bridge many gaps, from the amateur to professional astronomy, informal to formal astronomy educator, user of the telescope to the manufacturer and marketer of the telescope. Scott Roberts of Meade Instruments is one who comes to mind for the latter; many amateur astronomers know Scotty from his work at the various star parties and events around the country. But Scott steps way beyond with his incredible support of outreach.

Outreach also allows us to close one other gap: that of underrepresented populations in both amateur and professional astronomy. It gives us an opportunity to tell kids that regardless of who they are they can enjoy astronomy as a hobby or a profession. When I was the Executive Director of the Chabot Space & Science Center I often remarked that "I did not want to have every kid who came through those doors at Chabot and looked through a telescope become an astronomer, I just wanted to let them know they could do it if they wanted to!" And astronomy clubs have started bridging many of these gaps; despite the fact that amateur astronomy clubs and the outreach of club members and that of individual amateur astronomers still has a long way to go before a broad and diverse audience is served, they seem far better positioned to reach underrepresented populations. Progress has been made as involvement of underrepresented groups, particularly women, in astronomy clubs and societies in the United States is higher today than it used to be. We are not where we should be, but we seem to be heading in the right direction.

The Astronomical League, a federation of astronomy clubs and societies in the United States, has long been renowned for its Observation programs, which encourage member clubs to identify themselves as active in observing the night sky and which recognizes active observing amateur astronomers with rewards for certain levels of active observation. The concept of these Observing "Clubs" is to not just encourage amateurs to go out and observe the night sky, but to do so in a more formal way, recording their observations and receiving individual recognition for those observations. These Observing "Clubs" are rather popular with the amateur community, and include a number of clubs, such as Messier Club, Binocular Messier, Herschel 400 Club, Deep Sky Binocular Club, Meteor Club, Double Star Club, Lunar Club, Constellation Hunter Club, Southern Sky Telescopic Club, Arp Peculiar Galaxy Club, and the Asteroid Observing Club, to name just a few. There has been over 2,350 individual awards given for the Messier Club alone, dating back to 1967.

I was pondering how we could recognize individual outreach efforts when it dawned on me that a mechanism was already in place: the League's

Observing "Clubs"! Why not structure a League club that recognizes and rewards for individual outreach efforts? So I went to the League's Executive Council in 2005, made a proposal, and the next thing I knew I was the Coordinator of the Astronomical League Outreach Award. *Now that's what I get for opening my big mouth,* I thought to myself. But I knew this was important and hoped it would be successful.

When I conceptualized the Outreach Award, I did not think that many people would respond, even though the initial outreach level needed for a recognition or award of five 2-hour events was deemed as not too difficult to reach since the Astronomical League wanted to encourage and reward outreach efforts (see the appendix for the list of criteria). I was uncertain how many people were doing any sort of outreach, even though the Astronomical Society of the Pacific, through their Project ASTRO, and other programs such as the Night Sky Network had some basic data on the issue.

After my initial article introducing the new Outreach Award in the spring 2006 issue of the League's publication, *The Reflector,* submissions started rolling in. Emails were averaging two to ten a day, with all sorts of questions and, of course, that occasional phishing email. And by the end of year one well over 100 individuals had been awarded their League Outreach Award. As I reviewed the various submissions, their passion for outreach became very real to me. And the variety of activities—from the obvious, like telescope viewing, to extraordinary, like astronomical plays—was unexpected.

Now that the initial rush is over, we are seeing an average of about 6–8 submissions each month. A number of these are amateurs coming back for the next level. Some clubs and societies take this on as a club project; there are several clubs that have sent 10 or more submissions for the Outreach Award. To date there have been 120 awards presented, representing 3,600 hours of outreach at just over 1,000 events and a staggering 111,500 participants! I realize that some of these participants are double or even triple-counted and the participant counting mechanisms may not be statistically significant. Yet this gives one some idea of the dimension of education and public outreach (EPO) efforts. And I believe it is just the tip of the proverbial iceberg!

So just why do those of us who are not educators or employed at museums and sciences centers, where conducting "informal science education" is the mainstay, expend the effort to do astronomy outreach and the EPO effort? Is it to convince people to become astronomers, either amateur or professional? To boost memberships in local (or even national) astronomy or science clubs and societies? To sell astronomy and space books? Magazines? Telescopes, binoculars, and accessories? In the vast majority of cases, the answer is probably a resounding no… even from those who are trying to "sell" local clubs and society memberships, books, magazine, and telescopes; or maybe even a resounding "all of the above," depending on the individuals' motivation.

Exactly what is the bottom line, that is, the ultimate reason many of us find sharing the Universe is such a passion? I think that many of us simply love sharing the splendor and beauty of the Universe with others. We want others to see and experience what we have seen and experienced. We enjoy hearing the *ohs* and *ahs* as people view, perhaps for the first time, the majesty of our cratered Moon, the magnificence of Saturn and its rings, the gracefulness of Messier 42, or the Great Orion Nebula. In this society of "things" coming at people from all directions and a zeal—almost an expectation—for instant gratification, there is something about taking the time to look through an instrument whose technology is nearly 400 years old and seeing the grandeur of our Universe with one's own eyes. This is the real thing and not a view that one can capture by looking at a photograph. A picture might be worth a thousand words, but having those photons actually strike one's own eyes is an entirely different matter.

Some of the general public will never be smitten with what they see through our telescopes, and in some cases even quite disappointed. People have grown accustomed to seeing the spectacular images from the *Hubble Space Telescope, Mars Reconnaissance Orbiter, Cassini*, and *Galileo* spacecraft, to name just a few. The images, often "massaged" in Photoshop or similar processing programs to the nth degree, are not what one will ever see.

I vividly recall a fellow who contacted me in September 2006, eager to purchase a "good" (translated: expensive) telescope and was seeking my guidance. Money was no issue and he was willing to spend thousands for a telescope. His primary objective was "to take Hubble-like images." I tried to explain to him why that was not possible, unless he wanted to spend Kecklike money. (He didn't have quite that much to spend on his telescope.) So he went out and spent right at five figures on a state-of-the-art, high-quality telescope. After investing a lot of time showing him how to set up and use the telescope after it arrived, I did not hear from him for some time, so I gave him a call. "I decided those faint fuzzy things were not impressive," he said. "Not like the Hubble. So I returned the telescope and have moved on to something else." Not impressive? Was he looking at the same summer Milky Way objects I was seeing?

It can be quite mind boggling at times for people to ponder the meaning of what they might have seen through the eyepiece of a telescope or what they have been told about those objects. I have had people exclaim in disbelief about the delicacy of the rings of Saturn or the details in the mottled look of the Moon. People have expressed utter disbelief about the age of the light they saw from the Andromeda Galaxy. Yet they seem to come back to the eyepiece for a second gaze, and might even linger a little longer…

I emphasize observational astronomy in my introductory astronomy courses, and I am not surprised at the reaction of students to that which they see through the telescope or even a pair of binoculars. As I explain to my classes on the first night of Astronomy Lab that they will be *required* to attend at least one offsite dark sky observing session, I hear the groans and clearly see their body language which indicates: "this is not what I had in mind." But their reluctance quickly turns into enthusiasm when they look through the telescope or binoculars. At the end of the evening many ask, "When is the next observing session?" And each semester there are more students coming back for another public observing session, often bringing dates or their families. Where else can you promise the sun, moon, and stars and actually deliver?

The Fox Network started airing a new television reality program in spring 2007, entitled *Are You Smarter Than A Fifth Grader?* The host asks the contestants and a group of fifth graders who appear on each show a range of questions. These questions are arranged by topic and grade level, starting with first grade questions up to the fifth grade. As participants answer questions, they earn money, up to $1 million. After I heard about the program, I decided to watch one evening, curious about the contestants and whether astronomy questions would be asked. And indeed, there were astronomy questions; some of the questions include:

Of the following which kind of star is the hottest?

A. A blue giant

B. A red dwarf

C. A regular yellow star

Which planet is typically the brightest in the night sky?

In what constellation is the Big Dipper?

On the show's website a poll asks visitors which subject they know the least about? With competition like grammar and mathematics, astronomy still took top spot in this "least" poll and by quite some margin.

This is an indication that astronomy seems rather distant and untouchable for many. This is in part because of the way astronomy is taught and how astronomers are portrayed in popular culture: as scientists with extraordinary intelligence. The message that most people take away from this is: astronomy is not for me.

Amateur astronomers can overcome these barriers: They are regular people who were able to embrace astronomy and they present astronomy to others by focusing on the excitement and the phenomena, not the equations and the tests.

Many astronomy clubs and societies, as well as science museums and science centers, now include activities in their repertoire for conveying astronomy that make astronomy fun, engaging and exciting, and those activities go well beyond observing the night sky through a telescope.

Back to the telescope, that instrument amateur astronomers all use for a visual and personal connection to the universe. What is it about looking through a telescope or a pair of binoculars that turns on the light for many people, especially children? Is it the proverbial "taking the time to stop and smell the roses"? Is it exposure to something that seems so distant or untouchable? Is it that wonderful human characteristic of curiosity? Is it the thrill of experiencing authenticity, of one that is far away from us? Or is it something else entirely? No matter what it is, we ought to ask ourselves: What does this motivation mean for the kind of outreach that would or should be best done by amateur astronomers, and how can they be supported in their efforts?

Certainly programs aimed at supporting amateur astronomer outreach like those at the Astronomical Society of the Pacific (ASP) and the Astronomical League (AL) should be continued. The successes of the Night Sky Network and the League's Outreach Awards demonstrate interest by the amateur community. The fact that the National Science Foundation recognized the underlying philosophy of the Night Sky Network and now funds a new initiative to study and support amateur astronomers in their club-mediated outreach—as an example of the value that passionate hobbyists can provide to the public understanding of science—shows that amateur astronomers outreach could even provide a model for a lifelong learning society. Pairing up with the American Astronomical Society's professional astronomers and their public education efforts is sensible; many "pro-am" astronomical collaborations at colleges and observatories already exist. And the events and celebrations planned for 2009—the 400th anniversary of Galileo's first use of the telescope to explore the heavens—provide us with opportunities to build on these collaborations and perhaps capture more public attention.

Amateur astronomers are currently supported by the ASP and the AL, often with monies from NASA or the National Science Foundation. Beyond that, new initiatives are forming. A few private foundations have begun to spring up which have astronomy education and outreach as their core mission. The National Sharing the Skies Foundation, created by David and Wendee Levy, and the StarGarden Foundation, created by the late Vic Winter and Jen Dudley, are two such examples. They understand the power of outreach, from the beauty of an inspirational talk, to a hands-on demonstration, to a view through a telescope, and they support these kinds of activities.

Only once after viewing the Moon and Saturn through my telescope did I have someone tell me she thought it was all quite boring. Most people —children and adults alike—will appreciate the opportunity to see the splendor of the Universe. And remember that most of our audience will not become astronomers, either amateur or professional, and that is perfectly okay, for amateur astronomers have been given the opportunity to give them something that has enriched their lives and even their souls. And that, for many of us, is reward enough...

APPENDIX

Table 1. *Criteria for Receiving an Astronomical League Outreach Award*

Astronomy Outreach Award Level	Requirements—for each submitting individual

Outreach Award	• A minimum of five 2-hour (minimum each outreach) outreach events
	• Document each event:
	  ○ Date, time (started and ended), location
	  ○ What you did for the outreach
	  ○ Estimate of the number of people attending

Stellar Outreach Award	• In addition to the (basic) Outreach Award, the Stellar Outreach recipients will need an additional fifty hours (minimum) in outreach events
	• Again, document each event:
	  ○ Date, time (started and ended), location
	  ○ What you did for the outreach
	  ○ Estimate of the number of people attending
	• The recipient will "report" on one of his/her outreach events; these reports can be used in *The Reflector* and elsewhere to overview what amateurs are doing in Outreach and share ideas

Master Outreach Award	• In addition to the Outreach and Stellar Outreach Awards, the Master Outreach recipients will need an additional one hundred hours (minimum) in outreach events
	• As with the first two levels, document each event:
	  ○ Date, time (started and ended), location
	  ○ What you did for the outreach
	  ○ Estimate of the number of people attending
	• The Master Outreach Award nominee will report on what seems to work best for their outreach efforts; this can be specific activities, locations, etc. Like with the Stellar Outreach Award, these reports can be used in *The Reflector*

Biographies

Marni Berendsen. Berendsen, of the Astronomical Society of the Pacific, serves as the lead project manager of the Night Sky Network, a nationwide coalition of amateur astronomy clubs dedicated to astronomy outreach. An amateur astronomer for many years, she is a Project ASTRO partner and a member of the Mount Diablo Astronomical Society in Concord, California, participating regularly in the club's outreach programs and events. She received her Masters in Astronomy from the University of Western Sydney. Her interest in astronomy was sparked as a young Girl Scout camping out in a high Sierra meadow while Scout leaders told stories of the Greek gods illustrated by the tapestry of constellations over their heads.

Michael G. Gibbs, Ed.D. Gibbs is the Chief Advancement Officer for the Astronomical Society of the Pacific. Prior to serving at the ASP, he held various leadership positions at universities such as DePaul University in Chicago, Illinois. Gibbs received his B.A., M.S. and Ed.D. from DePaul. Gibbs also serves as an adjunct faculty member with the Spertus College Center for Nonprofit Management graduate program. He has lectured both nationally and internationally and coauthored several journal articles. Gibbs currently serves a member of the National Hispanic Institute's board of trustees.

James B. (Jim) Kaler, Ph.D. Kaler is Professor Emeritus of Astronomy and current President of the Astronomical Society of the Pacific, earned his A.B. at the University of Michigan, his Ph.D. at UCLA, and has been at the University of Illinois since 1964. His research area, in which he has published over 120 papers, involves dying stars. Professor Kaler has held Fulbright and Guggenheim Fellowships, has been awarded medals for his work from the University of Liège in Belgium and the University of Mexico, gave both the Armand Spitz Lecture to the Great Lakes Planetarium Association and the Margaret Noble Address to the Middle Atlantic Planetarium Society, and received the 2003 Campus Award for Excellence in Public Engagement. He has written for a variety of popular and semi-popular magazines, was a consultant for Time-Life Books on their Voyage Through the Universe series, has published several books, including *Stars and their Spectra, The Ever-Changing Sky,* and *Extreme Stars* (Cambridge), *Stars* and *Cosmic Clouds* (Scientific American Library), two textbooks, and *The Little Book of Stars* and *The Greatest Hundred Stars* (Copernicus). His latest book is *The Cambridge Encyclopedia of Stars.* Asteroid 1998 JK was named "17853 Kaler" in honor of his outreach activities.

Judy Koke. Koke is a Senior Research Associate at the Institute for Learning Innovation and has extensive background in free-choice learning research and evaluation. Previously the in-house evaluator at the Denver Museum of Nature and Science for six years, Koke was later the Assistant Director of the University of Colorado Museum of Natural History. She has published widely on evaluation findings and girls' attitudes towards science careers, and has taught in several graduate Museum Studies Programs. On the Board of the Visitor Studies Association, her research interests include public understanding of science, gender issues in science learning and learning in community-based organizations.

David H. Levy. David is tied for fourth place as one of the most successful comet discoverers in history. He has discovered 22 comets, 8 of them using his own backyard telescopes, the most recent in October 2006. With Eugene and Carolyn Shoemaker at the Palomar Observatory in California he discovered Shoemaker-Levy 9, the comet that collided with Jupiter in 1994. Levy is president of the National Sharing the Sky Foundation, dedicated to inspiring young people to be inspired by the night sky, and he is currently involved with the Jarnac Comet Survey, which is based at the

Jarnac Observatory in Vail, Arizona but which has telescopes planned for locations around the world. Levy is the author or editor of 35 books and other products. He won an Emmy in 1998 as part of the writing team for the Discovery Channel documentary, *Three Minutes to Impact* and is the Science Editor for *Parade* Magazine. A contributing editor for *Sky & Telescope* Magazine, he writes its monthly Star Trails column, and his Nightfall feature appears in each issue of the Canadian Magazine *Skynews*.Levy has given more than 1,000 lectures and major interviews, and has appeared on many television programs, such as the *Today* show, *Good Morning America*, the National Geographic special *Asteroids: Deadly Impact*, and ABC's *World News Tonight*, where he and the Shoemakers were named Persons of the Week for July 22, 1994. Also, Levy has done nationally broadcast testimonials for PBS (1995-present), and for the Muscular Dystrophy Association Telethon (1998-1999). He and his wife Wendee host a weekly radio show available worldwide at www.letstalkstars.com. In 2004 he was the Senator John Rhodes Chair in Public Policy and American Institutions at Arizona State University. He has been awarded four honorary doctorates, and asteroid 3673 (Levy) was named in his honor. Levy resides in Vail, Arizona, with his wife, Wendee.

Terry Mann. Mann is President of the Astronomical League, a JPL Solar System Ambassador and is an advisor for the Meade 4M Community. Mann received the G.R. Wright Award from the Astronomical League and the Hans Bauldauf Award for significant contributions to amateur astronomy. She has written articles for the Astronomical League's magazine, *The Reflector*, local newspapers, and her astrophotography has appeared in local art galleries, newspapers, and newscasts. As much as she enjoys observing, Mann has always devoted a large amount of her free time to education and public outreach. Mann is a frequent guest speaker and has lectured about astronomy at high schools, astronomy clubs and civic groups as well as science centers, and state parks.

Mike Reynolds, Ph.D. Reynolds is the Associate Dean of Mathematics & Natural Sciences and Professor of Astronomy at Florida Community

College in Jacksonville, Florida. Reynolds has 34 years in astronomy and space sciences experiencing a gamut of related professions (or roles), from a high school and university instructor to planetarium and museum director, researcher, writer, and lecturer. He received numerous recognitions for his work, including the 1986 Florida State Teacher of the Year, NASA Teacher-in-Space National Finalist, and the G. Bruce Blair Medal. Reynolds is the Executive Director of the Association of Lunar and Planetary Society and is the Executive Director Emeritus of the Chabot Space & Science Center in Oakland, California. He is a board member with the National Sharing the Skies Foundation, StarGarden Foundation, and the W Foundation. He is also on Meade Instruments 4M Board of Advisors. In 2003, Reynolds was invited to join the board of directors for the AstronomyOutreach network to promote and support outreach enthusiasts. He chairs the AstronomyOutreach Outreach Award committee to recognize individuals and organizations for exemplary achievements in awareness of astronomy to the public at large. Reynolds also designed and chairs the Astronomical League's Outreach Awards. These awards are given to individual League club/society members for their individual outreach efforts.

Scott W. Roberts. Roberts is a supporter of outreach in astronomy and space exploration, and a popularizer of amateur astronomy with over 20 years experience in the telescope manufacturing industry, and is currently a Vice President with telescope maker Meade Instruments. Roberts is the founder and Executive Director of the world's largest factory-sponsored organization devoted to outreach in astronomy, the Meade 4M Community. In 2000 Roberts created the AstronomyOutreach network that supports and recognizes astronomy EPO activists around the world, and sits on the board of the National Sharing the Sky Foundation. Roberts is also a JPL/NASA Solar System Ambassador. Since the early 1980s he has participated in education and public outreach in astronomy around the world through hands-on demonstrations, television, radio interviews, and print.

Timothy F. Slater, Ph.D. Slater is an associate professor of astronomy at the University of Arizona where his scholarship focuses on the teaching and learning of science. He is the Director of the Conceptual Astronomy and Physics Education Research (CAPER) group in the Astronomy Department and his research focuses on student conceptual understanding in formal and informal learning environments, inquiry-based curriculum development, and authentic assessment strategies, with a particular emphasis on non-science majors and pre-service teachers. Professor Slater earned his Ph.D. at the University of South Carolina in geophysics and his M.S. from Clemson University in astrophysics. He holds two bachelors' degrees from Kansas State University, one in science education and one in physical science. He has extensive experience working with teachers and students in Hawai'i and works closely with programs at UH Institute for Astronomy. Professor Slater is the elected education officer for the American Astronomical Society, an elected member of the Board of Directors for the Astronomical Society of the Pacific, serving as ASP Vice President, an elected councilor at large for the Society of College Science Teachers, is on the Editorial Board of the *Astronomy Education Review*, and has served multiple terms as chairman of the Astronomy Education Committee of the American Association of Physics Teachers. He represented the United States as the initial U.S. National Chairman of the 2009 International Year of Astronomy. He is an author on more than 70 refereed articles, winner of numerous awards, and is frequently an invited speaker on improving teaching of science through educational research and teacher education.

Martin Storksdieck, Ph.D. Storksdieck is a senior research associate at the Institute for Learning Innovation where his research interests include factors that influence cognitive and affective gains from free-choice learning experiences and determinants of behavioral and attitudinal change. Most of Storksdieck's research is centered in astronomy and environmental science, with some attention on current science and public understanding of the process and nature of science. Bridging the free-choice and the formal education sector, Martin's other interest lies in school field trips to a free-choice learning environment. Prior to joining the Institute, he worked as an environmental consultant specializing on local environmental management systems, was a science educator with a planetarium in Germany where he developed shows and programs on global environmental change, and served as editor, host, and producer for a weekly environmental news broadcast. He holds a Masters in Biology from Freiburg University (Germany), a Masters in Public Administration from Harvard University, and a Ph.D. in education from the University of Lüneburg (Germany).

Dan Zevin. Since 2002, Zevin has been the National Projects Manager and Project/Family ASTRO National Coordinator at the Astronomical Society of the Pacific (ASP) in San Francisco. For the most part, his job is to administer the day-to-day operations and overall logistics of national ASP programs currently or previously funded by the National Science Foundation. Prior to joining the ASP, Zevin was the Program Director at the Headlands Institute, a K–12 residential, outdoor science education provider in northern California. Before entering the education field, he worked for seven years as a wildlife biologist with the Nature Conservancy of Hawaii and at the Los Angeles Zoo. Zevin graduated from Humboldt State University, receiving his B.S. in wildlife management.